CONTENTS

THE KYOTO PROTOCOL: THE UNDERMINING OF AMERICAN PROSPERITY-THE SCIENCE

Wednesday, July 29, 1998

HOUSE OF REPRESENTATIVES,
COMMITTEE ON SMALL BUSINESS,
Washington, DC.

The Committee met, pursuant to notice, at 10:35 a.m., in room 2360, Rayburn House Office Building, Hon. Jim Talent [Chairman of the Committee] presiding.

Chairman TALENT. I call the Committee to order.

Today the Committee on Small Business will be holding its third hearing on the impact of the Kyoto Protocol, in which the United States agreed to reduce its production of greenhouse gases to below 1990 levels. Our previous hearings established that this is equivalent to raising the cost of energy to the American economy by 30 to 50 percent.

For Americans, this would mean: Gasoline prices would increase by as much as 65 cents per gallon; natural gas prices for industry would increase by 90 percent; the gross domestic product would decline by 2½ percent. The proponents of the Kyoto Protocol justify this effect on the American economy by the need to prevent what they describe as a catastrophic warming of the Earth's temperature.

I called this hearing with scientific witnesses to determine whether and to what extent the Earth is actually warming. My research into this subject suggests that the catastrophic global warming predictions of the Kyoto Protocol are based primarily on the pretense of a consensus within the scientific community, when in fact no such consensus exists; highly selective and misleading use of data; and failure to confront the body of evidence suggesting that the Earth's warming trend is natural and beneficial.

We have a number of distinguished scientists here today, several of whom will attempt to disabuse me of these views. The Committee is grateful for the time of these witnesses. I am especially grateful that several of the witnesses postponed important personal trips to be here today, and I am also grateful to all the witnesses for understanding why, in our grief over the death of Officers Chestnut and Gibson, the ranking member and I decided to postpone this hearing for several days.

I want to introduce the witnesses later. But right now I want to recognize our ranking member, the distinguished gentlelady from New York, Ms. Velazquez, for whatever opening comments she may wish to make.

Ms. VELAZQUEZ. Thank you, Mr. Chairman.

Today's hearing is another in a series that the Committee has held to examine the effects of global warming. Last month we examined the Kyoto Protocol and its effect on small business. Today we turn our attention to the signs of global warming.

While there may be question over how and in what way global warming exists, the effects of greenhouse gases on our environment are now becoming more and more evident. In the last century the Earth's temperature rose by 1 degree Fahrenheit. In fact, recent data not only shows that this is the warmest century on record, but also that the 1990's are shaping up to be the warmest decade ever.

The forecast is not any better. Scientists estimate that over the next century temperatures will continue to rise another 1.8 to 6 degrees. We already are witnessing severe weather patterns occurring with greater frequency. The resulting damage to our ecosystem has drastically escalated. Shifts in our agricultural growth conditions are beginning to rob regions of their ability to produce food.

The threat to human health has also increased. There is a reduction in the availability of fresh water. We have also seen an increase in the range and instances of disease. Estimates show that by the year 2100 we may see $50 to $80 million more cases of malaria alone.

Today's hearing on the science behind Kyoto will hopefully shed some light on how global warming will affect regional weather patterns and the rate of climate change. Furthermore, I hope we will hear what effect higher temperatures will have on health, the ecosystem, and socioeconomic conditions. Today's hearing should also provide some insight into possible solutions to address these problems.

Let me close by once again thanking the Chairman, Mr. Talent. Thank you.

Chairman TALENT. I appreciate the comments of the gentlelady.

I will introduce each witness as he testifies. The first witness today is Dr. Robert T. Watson, who is the Chairman of the Intergovernmental Panel on Climate Change. He resides in Washington, DC.

Dr. Watson, I want to thank you for being here today, and especially for postponing a trip that you had planned in order to attend. It was very kind of you. Please proceed with your testimony.

STATEMENT OF ROBERT T. WATSON, CHAIRMAN, INTERGOVERNMENTAL PANEL ON CLIMATE CHANGE

Mr. WATSON. Thank you. Mr. Chairman, Members of the Committee, it is a pleasure to be here today to discuss what is really a very important issue for both this and future generations. I am testifying in my capacity as the new Chair of the Intergovernmental Panel on Climate Change.

Despite what you will hear from several members of this particular panel, there is in my opinion no doubt that the overwhelming majority of scientific experts, while recognizing that scientific uncertainties do exist, still believe that human-induced climate change is inevitable. There are a number of questions, however: What will be the rate of climate change, the magnitude of climate change, and where specifically will climate change occur?

THE KYOTO PROTOCOL: THE UNDERMINING OF AMERICAN PROSPERITY-THE SCIENCE

HEARING

BEFORE THE

COMMITTEE ON SMALL BUSINESS HOUSE OF REPRESENTATIVES

ONE HUNDRED FIFTH CONGRESS

SECOND SESSION

WASHINGTON, DC, JULY 29, 1998

Printed for the use of the Committee on Small Business

Serial No. 105–62

U.S. GOVERNMENT PRINTING OFFICE

50–206 CC WASHINGTON : 1998

For sale by the U.S. Government Printing Office
Superintendent of Documents, Congressional Sales Office, Washington, DC 20402
ISBN 0-16-057559-1

COMMITTEE ON SMALL BUSINESS

JAMES M. TALENT, Missouri, *Chairman*

LARRY COMBEST, Texas
JOEL HEFLEY, Colorado
DONALD A. MANZULLO, Illinois
ROSCOE G. BARTLETT, Maryland
LINDA SMITH, Washington
FRANK A. LoBIONDO, New Jersey
SUE W. KELLY, New York
MARK E. SOUDER, Indiana
STEVEN J. CHABOT, Ohio
JIM RYUN, Kansas
VINCE SNOWBARGER, Kansas
MICHAEL PAPPAS, New Jersey
PHIL ENGLISH, Pennsylvania
DAVID M. McINTOSH, Indiana
JO ANN EMERSON, Missouri
RICK HILL, Montana
JOHN SUNUNU, New Hampshire
JOSEPH R. PITTS, Pennsylvania

NYDIA M. VELÁZQUEZ, New York
JOHN J. LaFALCE, New York
NORMAN SISISKY, Virginia
GLENN POSHARD, Illinois
JESSE JACKSON, Jr., Illinois
JUANITA MILLENDER-McDONALD, California
DANNY K. DAVIS, Illinois
ALLEN BOYD, Jr., Florida
CAROLYN McCARTHY, New York
BILL PASCRELL, Jr., New Jersey
RUBEN HINOJOSA, Texas
DONNA M. CHRISTIAN-GREEN, Virgin Islands
ROBERT A. BRADY, Pennsylvania
VACANCY
VACANCY
VACANCY

Mary McKenzie, *Staff Director*
Michael Day, *Minority Staff Director*

The large majority of scientists also believe that, on balance, there will be a number of adverse effects: human health, heat-stress mortality and increase in vector-borne diseases such as malaria; changes in ecological systems, particularly forested systems and coral reefs; and, indeed, changes in socioeconomic sectors such as agriculture, forestry, fisheries, water resources, and human settlements.

What I will testify today is not my own work but indeed the careful work of thousands of experts: climatologists, modelers, oceanographers, meteorologists, chemists, biologists, technologists, economists, and social scientists. These experts come from universities, government laboratories, the private sector, and from industry.

What decisionmakers should take into account is that the atmospheric residence time of carbon dioxide is extremely long, more than a century; and hence, if indeed policy formulation waits until all scientific uncertainties are resolved, and if indeed greenhouse gases are responsible for changing the Earth's climate, as projected by all climate models, the time to reverse that damage would actually be centuries to millennia.

I have a Figure 1 which actually shows that if you doubled atmospheric CO2 over the next 70 years in the atmosphere and then kept it absolutely level for the next 1,000 years, sea level would continue to increase almost monotonically for the next 600 years and a long time thereafter.

So a key question for policymakers is, what level of information do you need to make policy? A decision to act or not to act will indeed have profound implications.

Simply stated, however, greenhouse gases have increased in the atmosphere due to human activities. There is no doubt that carbon dioxide has increased by nearly 30 percent since the preindustrial era, the mid 1880's; and indeed this is due largely to changes in land use, deforestation, and due to the use of fossil fuels, coal, oil, and gas.

The Earth's atmosphere has indeed warmed over the last 100 years about half a degree centigrade, slightly more on land, slightly less in the oceans. The 1980's and the 1990's are shaping up to be the two hottest decades of the last 600 years. Precipitation changes are occurring. Tom Karl's work at NOAA has shown there were changes in both the spatial patterns and the temporal patterns of precipitation, with more precipitation falling in heavy precipitation events.

The theoretical models, when we try to compare them to the observations, are in fairly good agreement between theory and observation in spite of what you will hear from several other witnesses. If you take into account all the changes occurring, that is to say, changes in greenhouse gases and aerosols, there is indeed reasonable agreement in the Max Planck model, the U.S. Met Office model, and the GFLD model in Princeton, there is reasonable agreement between theory and observations.

If we take into account what is likely to occur over the next 100 years due to economic growth, population growth, one will expect a significant increase in the emissions of greenhouse gases such as carbon dioxide. This is projected to lead to a temperature increase of somewhere between 1 and 3½ degrees centigrade by the year 2100.

Indeed, that is a very wide range. As Pat Michaels points out quite correctly, there are a significant number of uncertainties. That is why we do not take a single point. That is why we take a range.

The question is, why should we care? I have already alluded to it. Not only will there be changes in temperature but changes in precipitation patterns and sea level. We project that sea level will also increase by between 15 and 95 centimeters by the year 2100.

What are the implications? As I have already said, there will be implications for natural ecological systems, especially forests and coral reefs, as a sustained temperature increase in the ocean will clearly have adverse effects on the viability of many coral reefs around the world. Forested systems will also be threatened, even after you take into account the positive effects of carbon dioxide fertilization. We go into some detail of this in the IPCC report.

With respect to food, we believe that globally there may not be any adverse effects of climate change on global food production, but a significant decrease in the tropics and subtropics where people are already hungry today, and an increase actually in production in the high northern latitudes of the USA, Canada, and Siberia. This takes into account the positive CO2 effect. If we did not take the positive CO2 effect into account, we would actually project a decrease in productivity globally. So we do take into account what is mentioned by some of the other witnesses, the so-called beneficial effects of carbon dioxide.

We have to recognize that more than a billion people across the world today do not have adequate supplies of fresh water. Nineteen countries today are water scarce or water stressed, many of them in Africa and the Middle East. Most of the projections would suggest that these arid and semiarid areas will become even more arid and semiarid in the future under the climate change scenarios that we project.

Sea level is projected to increase, as I stated earlier, 15 to 95 centimeters. This could lead to a significant loss of land in lowland deltaic areas in Egypt, Bangladesh, and China, with tens of millions of people being displaced if we do not build seawalls. Countries such as Bangladesh do not have the financial or institutional infrastructure to take care of their defense against sea level rise. In addition, many small island states will be completely threatened.

The issues on human health are indeed vexing in diseases such as malaria. There are currently 200 million cases of malaria per year. Two million people die per year from malaria, primarily in the tropics. There would be a projected increase in the incidence of malaria, once again primarily in places where there are not good health care facilities. This is not a major threat to the United States, but it is a threat to places such as Africa.

What are the social costs of inaction? There is a range from the IPCC of 1 to 2½ percent of gross domestic product across the world, with a number of 5 to 9 percent for developing countries. These numbers are very, very soft and very squishy. But the key point is there is a cost of inaction. Unfortunately, developing countries are much more vulnerable to climate change than developed countries.

The key question is, effectively, at what level does it need to stabilize? Many people keep saying to stabilize at 1990 levels of carbon dioxide would require a 60 to 80 percent cutback in CO2 emissions. That is correct. However, if indeed one wants to stabilize at 550 parts per million, a level suggested by the European Union, that does not take draconian action today. It needs careful, planned action over the next 50 to 100 years.

Indeed, if one looks at some companies such as Shell and British Petroleum, they put a plausible future forward where by 2050, in a very logical and progressive manner, half of the world's energy would be produced from renewable energy, and this is a scenario that stabilizes around 550 parts per million without severe economic dislocation. This is some work that has come out of Shell and BP, and they are looking to see basically how to look at Kyoto as an opportunity rather than just a basic problem.

So in conclusion, Mr. Chairman, Members of the Committee, the earth's climate is changing. We cannot explain it on natural variability alone. We project climate will change significantly in the future, with potential adverse effects on human health, ecological systems and socioeconomic systems; that there are a number of measures that can be taken that can cost-effectively address this challenge.

Thank you.

[Mr. Watson's statement may be found in the appendix.]

Chairman TALENT. Thank you, Dr. Watson, and I appreciate your summarizing your written statement. I normally ask the witnesses, and forgot to do so, to summarize their statements if they can to leave enough of time for questions.

I am also going to say to the Members, and we are going to experiment here. If you have a clarifying question, a lot of this material gets technical, and if you do not understand a statement made by somebody and you want to clarify it while it is fresh in your mind and in everybody's mind, I will allow some limited questions during the testimony if we can manage that.

Now, if it gets into long exchanges, I may have to go back to the normal rule. So if one of the witnesses says something — and I hope you all will appreciate that we may interrupt you to say "what do you mean when you talk about this climate and what are you saying about that" — then feel free to seek recognition for a question or two just to clarify. If you are trying to make a point, then I would appreciate it if you would wait until your regular question period.

Our next witness is Dr. Daniel A. Lashof, who is the senior scientist for the Natural Resources Defense Council.

Dr. Lashof.

STATEMENT OF DANIEL A. LASHOF, SENIOR SCIENTIST, NATURAL RESOURCES DEFENSE COUNCIL

Mr. LASHOF. Thank you very much, Mr. Chairman, Members of the Committee.

It is an indisputable fact that atmospheric concentrations of greenhouse gases are climbing and that that rising trend can be attributed to human activities. The concentration of carbon dioxide in the atmosphere today globally is over 360 parts per million, a level that is substantially higher than any concentration that has been experienced for at least 200,000 years of the Earth's history.

The most detailed scientific assessment of an environmental problem ever conducted, the Second Assessment Report of the Intergovernmental Panel on Climate Change just referred to by Dr. Watson, concluded that global warming is already upon us and that, unless emissions of greenhouse gases are reduced significantly, future impacts on human health and the environment are likely to be severe.

Let me give just a couple of examples of the predicted effects in recent experience. The IPCC projects that with global warming there will be more frequent and intense heat waves. Recent experience is compelling. Each month so far this year has set an all-time record for global near-surface temperatures since instrumental records began over 100 years ago. June 1998 exceeded the previous record warm June, which occurred in 1994, by nearly .4 degrees Fahrenheit. Now while that may not sound like a lot, as a global average change in temperature it is unprecedented.

The current heat wave, which has claimed tragically more than a hundred lives in the United States this year, as well as the heat wave that killed more than 400 people in Chicago in 1995, are examples of the kind of event that are projected to become more and more frequent in the decades ahead.

According to NOAA, in Brownsville, Texas this year they recorded 17 days in June that had a minimum temperature at or above 80 degrees Fahrenheit. This would have a probability of recurring in a stable climate only once in a thousand years. But by the middle of the next century this is expected to be a 1 in 3-year event if greenhouse gas pollution is not curtailed.

Another example: In Melbourne, Florida there were 24 days with temperatures above 95 degrees in June. This is also an unprecedented event with a probability of recurrence of once in a thousand years.

Extreme drought and more severe storms in our heartland are also projected. More of our rainfall is now coming in intense downpours rather than steady drizzles, as Dr. Watson mentioned. This, coupled with higher rates of evaporation in a warmer world, is expected to intensify droughts and storm damage in America's breadbasket.

The current drought in the southeastern United States is an example of this kind of event. Last winter was very wet and soil moisture levels, as illustrated in my testimony, were well above normal in the winter. Yet severe drought conditions developed very rapidly in the spring when the normal spring rainfall did not occur, and now soil moisture is dramatically below normal levels, which had a direct effect in allowing the wildfires to take place in Florida which consumed more than half a million acres in June.

The drought in the southeast is causing severe damage. In Texas $1.5 billion in agricultural losses are expected, over $2 billion of losses in Oklahoma, $175 million of losses in Florida and $400 million in losses in Georgia.

Mr. Chairman, it is impossible to unequivocally demonstrate a link between global warming and any specific event. The examples I have given, however, demonstrate that global warming can have severe consequences. It is abundantly clear that we are facing a real and serious threat.

Now, the climate system is complex and there are many remaining uncertainties, as Dr. Watson has indicated, particularly about the details of climate change at the regional and local levels. As we discuss these uncertainties, it is essential to remember that uncertainty is a double-edged sword.

Pat Michaels, in his written testimony, confidently predicts that global warming forecasts will be revised downward in the future, and I cannot rule that out. That is a possibility. But as the history of the ozone depletion problem demonstrates, it is just as likely that things will turn out worse than the current mainstream forecast.

For example, a recent article in the Journal Nature reviews the risk that sea level could rise by as much as 20 feet over the next several centuries, rather than the mainstream forecast of 1 to 3 feet, as a result of a rapid collapse in the west Antarctic ice sheet. In the most recent issue of Science, recent measurements are reported which heighten this concern, indicating that a key Antarctic glacier is retreating at a surprisingly rapid rate of 1.2 kilometers per year.

Now, Mr. Chairman, I do not suggest that you rush to judgment based on these recent papers. Rather, they need to be assessed, along with other findings that are reported in the literature literally every week, before they become the basis for policy. The same is true of the papers cited by Pat Michaels.

That is the reason why the Bush Administration established and supported the Intergovernmental Panel on Climate Change, and why that support has been continued by subsequent administrations. We need a rational process of assessing all the data as a basis for policy.

To emphasize this point, let me give just two examples of how selective use of the recent literature can be very misleading. In his written testimony, Pat refers to a recent paper by Jim Hansen of NASA to suggest the current rate of increase in carbon dioxide is "below the most conservative U.N. scenarios." Unfortunately, this statement is simply false.

The main scenario used in the IPCC Second Assessment Report predicts that CO2 concentrations should be increasing by about half a percent per year in the decade of the 1990's, and the observed rate is about .4 percent per year, which is insignificantly different given the variability of the record over a short period of time.

Now, it is true that some papers that were published many years ago had assumed higher rates of change, but the IPCC scenario that Dr. Watson and Dr. Michaels referred to does not. They use more up-to-date data. But since Pat has cited Hansen's paper, he presumably believes that that is a credible source of information on this issue, so I would like to read from part of the paper that he has omitted from his testimony. I quote here.

"We propose an index of climate change based on practical climate indicators such as heating degree days and the frequency of intense precipitation. We find that in most regions this index is positive, the sense predicted to accompany global warming. In a few regions, especially in Asia and western North America, the index indicates the climate change should be apparent already, but in most cases climate changes are too small to stand out above year-to-year variability. The climate index is strongly correlated with global surface temperature, which has increased as rapidly as projected by climate models in the 1980's. We argue that the global area with obvious climate change will increase notably in the next few years."

My second example of unfortunate selective use of information, Mr. Chairman, actually comes from a letter that you circulated on July 16; and I am afraid that you have been quite poorly advised on the basis of that letter. It refers to a paper by Keigwin that was published in the journal Science, and I have that paper here. That is a very interesting paper and it is a useful contribution to the scientific literature. However, I would like to point out that it is based on a single sediment core in the Sargasso Sea. You use that paper to suggest that recent warming is a recovery from the Little Ice Age.

However, the IPCC assessment and a recent paper in Nature by Mann, et al., used a much more comprehensive assessment of all the available indicators of past climate change over the past 600 years and conclude, with 99-percent statistical confidence, that 1997 was the hottest year in the Northern Hemisphere in the last 600 years. There simply is not enough data for the Southern Hemisphere to make a similar calculation.

Mr. Chairman, in conclusion, the Defense Department would never base its response to military threats on the views of a handful of doubters, on the hope that the world will turn out to be benign. We should apply the same precautionary principle to global warming policy.

Indeed, as a party to the United Nations Framework Convention on Climate Change, signed by President Bush and ratified by the Senate in 1992, the United States is officially committed to this precautionary approach. The 1992 convention sets the objective of preventing dangerous anthropogenic interference with the climate system. Achieving this objective requires substantial and progressive reductions in global greenhouse emissions beginning without delay.

Thank you.

[Mr. Lashof's statement may be found in the appendix.]

Chairman TALENT. Thank you, Dr. Lashof.

Our next witness is Dr. Patrick Michaels, the Senior Fellow in Environmental Studies for the Cato Institute in Washington DC.

Dr. Michaels.

STATEMENT OF PATRICK J. MICHAELS, SENIOR FELLOW IN ENVIRONMENTAL STUDIES, CATO INSTITUTE

Mr. MICHAELS. Thank you.

If you do not mind, I would like to use some slides, so I am going to move over to where the projector is.

Chairman TALENT. That would be fine. When I was in science class, I always appreciated when they had movies or slides. So this will be a welcome change for the Committee.

Mr. MICHAELS. 10 years ago, almost exactly, in June 1988, NASA scientist James Hansen testified before the House that there in was a strong "cause and effect relationship" between the current climate, which was then in a drought and heat wave much worse than the summer of 1998, and human alteration of the atmosphere.

At that time, he had produced a model of climate which was published in the Journal of Geophysics, and the model projected the temperature increases for the next 10 years and further on. That model was very similar to five other climate models that served as the basis for this statement from the United Nations Intergovernmental Panel on Climate Change in 1990, which said that "when the latest atmospheric models are run with the current concentrations of greenhouse gases, the simulation of present climate is generally realistic on large scales."

The projection in Hansen's model, which actually included years, was the climate change from 1988 to 1997 would be .45 degrees Celsius. The observed changes from 1988 to 1997 are given here along with that projected increase of .45 degrees Celsius. The open-circle record is the surface temperature record primarily from land areas, primarily from the middle latitude, with very little in the high latitudes. The bottom two records are the satellite-measured temperatures, which are truly global, of the lower troposphere, and the balloon-measured temperatures, which actually are more global than the surface record because of launches from some very remote locations.

As you can see, the forecast that was made in 1988 was a failure; and the statement in 1990 that the type of models that were used then, which were the models that served as the basis for the Framework Convention on Climate Change and the subsequent Kyoto Protocol, were generally realistic was simply not true.

Chairman TALENT. Dr. Michaels, start with the top line and go down through the four lines and identify them again.

One of the things I caution the witnesses is not to assume a familiarity with anything on the level that you all are familiar with. I am trying in a very delicate way to say that you have to take this slow, at least for me and I suspect for other Members of the Committee, but I will take the blame for it. So go from the top to the bottom and start with that solid line at the top and tell us what that is.

Mr. MICHAELS. The solid line is the projected increase from the 1988 model that Hansen published. It was used as the basis for his statements about the future.

The middle line is the surface measured temperature by thermometers all around the planet; that is the Intergovernmental Panel on Climate changes, a record. The net change in that record, by the way, is a warming of .11 degrees over the past 10 years, when the forecast was .45 degrees. That is a factor of 4 degrees beneath where it is supposed to be.

The other two records are satellite-measured temperatures that John Christy is going to talk about; they measure the temperature of the lower atmosphere. The triangles are weather balloon-measured temperatures in the same zone that the satellites sense, and it is really quite remarkable to see the agreement between the two. Those two records actually show declines in temperature over the course of the decade that was supposed to have warmed .45 degrees.

Ten years ago I was asked to testify in front of the House and I made the observation that the current models which were extant, like this, predicted that the temperature of the planet should have warmed about 1.5 degrees over the course of the 20th century, and the observed warming, as Bob Watson said, was 0.5 degrees Celsius. So that, the error that these models have been making, has been remarkably consistent through the course of the 1990's, the 1980's, and the 1900's.

That led to what is generally thought to be sort of the paradigm of moderate climate change, which is very simply stated. If the models are consistently making an error producing about one-third as much warming as there has been, — and these models have since been modified, and we will touch on that in the Q and A, I am sure — then the upcoming warming should be about one-third of what they are predicting; and that would work out to be 1 to 1.5 degrees of Celsius.

Well, some remarkable things have happened in the last 6 months that give us reason to believe that we now understand why there was such a discrepancy between what was being forecast and what was being observed, and I would like to talk about those three important things that have appeared in the scientific literature.

Most important is the fact that the carbon dioxide increase in the atmosphere is not going in at the rate that was assumed by most of the models. Most models assumed a 1 percent increase in carbon dioxide per year. The increase has been far below that.

Second, methane in the atmosphere is the second most important of the greenhouse gases, the gases that have the potential, actually the reality of warming the lower layers of the planet. According to Dlugokencky, I believe that is the correct pronouncement, in Nature magazine, the rate of methane increase began to decline in 1981, 17 years ago, and is now coming into equilibrium with the other chemical reactants related to the atmosphere, so that very little increase in methane is projected now over the course of the next century.

Third, and perhaps most astounding, is a report by Myhre, et al. that shows that we overestimated the direct warming effect of carbon dioxide itself by 15 percent.

When you take all these factors into account and look at the past history of this issue, a very interesting picture emerges. In the first IPCC report in 1990, a median warming of 3.2 degrees was predicted by the IPCC in that first assessment. In 1992, in their report which was prepared specifically for the Framework Convention Meeting in Rio, the warming dropped to 2.6 degrees, this mainly because of the Montreal Protocol which reduced chlorofluorocarbons. In 1995, it dropped 2.0 degrees, their most likely value, given the executive summary of the 1995 report.

Now, if we allow for an additional four-tenths of a degree drop because CO2 is not going in as fast as it was assumed to, and recognize that methane is supposed to be responsible for about 15 percent of the warming of the 21st century, if methane does not increase you have to drop the forecast by .15 times 2 degrees, or .3 degrees, and recognize that the carbon dioxide radiative effect directly was overestimated by 15 percent, that drops it by another three-tenths of a degree.

We just got to one degree. This is the value that has generally been given, based upon the data, and it leads one to ask the question, how much trouble might we have saved if we had listened to the data rather than the computer?

Chairman TALENT. Put that back just a second if you would, doctor. I think, if I understand this correctly, I want to make clear to the Members, the first three bars on that graph are what the IPCC has projected. But that last one, I do not want Members to think that that 1.0 is an IPCC projection at this point; that is, as I understand it, your judgment of what they should do in light of the changes that you have been talking about. I just want to make sure that Members understand that it is not what the IPCC has done to this point. That is correct, right?

Mr. MICHAELS. Yes. I believe that the median value that they will project in the future will have to be lower than their current median value. Whether it will come all the way down to 1.0 degrees, I do not know the answer to that. I suspect it will not.

Chairman TALENT. We have the IPCC here, in a manner of speaking.

Mr. MICHAELS. He cannot speak for everybody. There are very large and vociferous arguments that take place, and Bob will be happy to enlighten us about those, I am sure.

Well, anyway, the question then becomes, what has happened over the course of the last 100 years? It is rather misleading to go around saying that the hot temperatures of the summer are a result of global warming. By the way, the summer in the United States this year is not all that out of line.

What global warming does, when you change the greenhouse effect, is it primarily warms the driest air masses on the planet. That is the physics of the greenhouse effect. This is the temperature change that you see since World War II.

I do not understand what Dr. Watson said, that there are some scientists here who do not believe human beings are changing the climate. I do not think there is anybody at the table who does not think that this change is awfully suspicious. What you see, though, is the warming is almost all in the winter. This is the winter on the planet. I also flipped the equator here, so it is winter down here.

The major warming is here in Siberia, from about –40 in the coldest air masses to about –38 in northwestern North America. If you take these winter warmings out of this picture, this map becomes very, very un-pink, with a mean temperature somewhere around there.

Now, let's take a look at the change in the summer temperature, where the air masses are moister, except for around the Sahara Desert, and therefore the greenhouse effect from carbon dioxide does not influence the temperature as much. This was published in the Journal of Climate Research about 2 months ago, by the way.

This is the change in the summer temperature, and this map is decidedly less bright red than the temperature change map for the winter.

Well, another thing that we heard earlier today was that variability, extreme events, are increasing. It is easy to do that. You can take one variable out of the plethora, the universe of weather variables, and find increases in it. You can take, on the other hand, say, regional temperature around the globe and look at the change between seasons and between years, which will be a true measure as to is the weather becoming more extreme or is it not.

This is from the last issue of the journal Climate Research, and what you see is a statistically significant decline in variability. This should surprise no one, given the fact that it is the coldest temperatures of the winter that are showing the warming, because winter is a much more variable time than the rest of the year, so if you warm up the winter what you do is reduce the variability.

I would like to finish with a little bit of discussion about the Kyoto Protocol. That is why we are here. The first question I would like to ask is this: If the United Nations had said in 1990 that the net warming of the 21st century would be between 1 and 1½ degrees Celsius, would we have the Framework Convention on Climate Change?

That is an important question I cannot answer. But I think I can answer, I think negatively, the following question: We would not have the Kyoto Protocol to that Convention, but given the fact that we do have the Kyoto Protocol, let's ask what it will do to the climate.

We will ask Tom Wigley from the United States National Center for Atmospheric Research, a senior scientist, who just published a paper which assumes that every nation on the planet does what it was supposed to according to the Kyoto Protocol. The temperature change that results from every nation agreeing to do what it does is 0.07 degrees Celsius. That is right, a number so small it could not be detected by any system that we have, but a cost that would be enormous. This is all cost and no benefit.

If we assume that the warming will be less, 1 to 1.5 degrees, which I think is warranted by the data and now by the physical science that underlays the data, you get a change of temperature of .04 degrees, four-hundredths of a degree of temperature saved by the year 2050 according to the Kyoto Protocol. Those are numbers coming out of the National Center for Atmospheric Research.

In conclusion, I think we again have to ask the question: Given these numbers, what should we do? how much should we spend? Is the Kyoto Protocol worth the cost?

Thank you very much.

[Mr. Michaels' statement may be found in the appendix.]

Mr. BARTLETT. [Presiding.] Thank you very much for your testimony. Our Chairman has had to leave for a vote on a markup in another Committee.

Our next witness is Dr. Fred Singer, President of Science and Environmental Policy Project, Fairfax, Virginia.

STATEMENT OF S. FRED SINGER, PRESIDENT, THE SCIENCE & ENVIRONMENTAL POLICY PROJECT

Mr. SINGER. Thank you, Mr. Chairman and Members of the Committee.

I have a very simple message, which is that climate science does not support the Kyoto Protocol and the emission controls on carbon dioxide. As will become evident from my testimony, the climate is not currently warming. The climate models used to predict the future warming have not been validated by actual observations.

Second, the UN-sponsored scientific report by the Intergovernmental Panel on Climate Change that you heard about, the IPCC, has been altered to convey the misleading impression to the public that there is a discernible human influence on climate which will lead to catastrophic warming.

Finally, as I will show, the Climate Treaty itself does not have a defined goal. It does not provide any guidance of what level of greenhouse gases to aim for. So the decisions of the Kyoto Protocol to cut emissions by 5.2 percent are arbitrary, purely political, and not in any way science-based. There is no scientific backup for these numbers.

My scientific background is described in the written testimony. I will not go over it here, except to mention that I published a book last year which reviews carefully the published literature in the field, the work done by many scientists, most of them IPCC scientists, the same scientists who also worked on the IPCC report. But my conclusion, after reviewing their work, does not agree with the summary of the IPCC report. I come to a different conclusion.

I can encapsulate my points here in five ways. Let me do point first.

The history of the Earth shows many natural climate fluctuations. We have accurate data going back for 200,000 years that does not support any of the claims that warmer temperatures would be catastrophic. We have studies from tree rings, ice cores, and ocean sediment cores. All indicate that the climate was much warmer than today 1,000 years ago, 3,000 years ago and 6,000 years ago; that these temperature changes were as large or as rapid as those being forecast by climate models; that they have already occurred and there were no detrimental effects that we know of.

Humanity and civilizations survived all of these natural global warmings of the past with far fewer resources than we have today, and they thrived during the warm periods. We know that from recorded human history. Life was better a thousand years ago during the Medieval Climate Optimum than it was during the Middle Ice Age 200 years ago.

Ms. MILLENDER-MCDONALD. Mr. Chairman, may I ask a question?

Mr. BARTLETT. Excuse me, Dr. Singer. There is a question.

Ms. MILLENDER-MCDONALD. Dr. Singer, you just mentioned that the climate was much warmer 1,000 years ago, 3,000, 6,000 years ago. What was the population at that time?

Mr. SINGER. Much smaller than it is today. Much less. Much poorer.

Ms. MILLENDER-MCDONALD. But much smaller.

Mr. SINGER. Smaller and poorer, yes.

Ms. MILLENDER-MCDONALD. Absolutely. Thank you so much.

Mr. SINGER. However, there were large concentrations of people in cities at the time, as there are today. But, overall, the population level was much less.

My second point: The theoretic models that are used to predict a future warming that you have heard about, these models have not been validated by actual atmospheric data. They need to be validated so that we can believe in them. The present models cannot handle clouds very well. I have slides I can show that different models handle clouds in different ways with different outcomes.

Other important climate factors are also not well-described by these models. The reason is that the atmosphere is very complicated and the models are still very crude.

Now, the actual data we have, the surface data that you have seen, that Pat Michaels projected, that shows light warming, are actually contaminated by local urban effects. Furthermore, the surface data cover only a very small fraction of the globe.

The only good data we have covering the whole globe that are accurate are satellite data, because satellites go around the whole globe daily and take accurate measurements; and these satellite data have not shown a warming trend since records were kept since 20 years ago.

My third point deals with the impacts. We have solid evidence, and agriculturalists all agree, that the ongoing increase in atmospheric carbon dioxide speeds up the growth of plants. It is good for agriculture. The empirical evidence indicates that a modest climate warming, from whatever cause, would actually lower sea level, not raise it. This is an important point and contradicts the stories that you have been hearing about. The evidence is here. I will be happy to show it in detail if I am asked about this.

What one should fear most about climate change, in my view, is a return of the next Ice Age. That is inevitable. It is bound to come. We do not quite know when. It is now overdue, because we have been in a warm period between Ice Ages for over 10,000 years and these warm periods usually do not last more than 10,000 years.

Fourth: This is an important legal point. The Climate Treaty, in Article II, clearly states the goal of the treaty. The goal of the treaty is not to lower emissions of greenhouse gases. The goal of the treaty, to quote, is to "achieve stabilization of greenhouse gas concentrations in the atmosphere at a level that would prevent dangerous anthropogenic interference with the climate system."

No one knows what this means. Nobody knows what constitutes a dangerous concentration of greenhouse gases that would cause problems for the climate system. There is no scientific basis for defining such a confrontation. We do not even know whether more or less carbon dioxide in the atmosphere would produce a dangerous concentration.

We do know that if we wish to stabilize carbon dioxide at the present levels, according to the IPCC data, we would have to reduce emissions by 60 to 80 percent worldwide. We would have to cut energy use worldwide by 60 to 80 percent to stabilize carbon dioxide at the present level.

This policy would severely damage not only our own economy, but it would foreclose economic development for most of the world's populations and condemn them to poverty. You cannot improve your standard of living without energy.

Fifth, my final point, is that there is no scientific consensus that global warming is occurring as a result of human activities. Some 17,000 scientists, over 2,300 of them specialists in fields related to global warming, have signed a petition against the Kyoto Protocol, the Oregon Petition; 17,000 scientists who have no particular axe to grind.

Despite what you may have been led to believe, the scientists contributing to the IPCC science reports have never been surveyed or polled as to their views, and the IPCC report summaries themselves are ambiguous and inconclusive.

In conclusion, despite the fact that science does not support the emission cuts called for in the Kyoto Protocol, I personally do not argue for complacency. Any human-induced change in environment must be carefully monitored and evaluated.

In the meantime, however, what should we do? I believe that common sense, no-regrets policies like cost-effective energy conservation and improved efficiency are in order, rather than hasty and economically damaging actions to cut energy use based on insufficient science.

Thank you.

[Mr. Singer's statement may be found in the appendix.]

Mr. BARTLETT. Thank you very much.

As you heard during your testimony, the bells have gone off for a vote. Fortunately, it is only a single vote, so we will be able to return very shortly from that. So the Committee will be in recess very briefly for a quick vote and then we shall return. Thank you.

[Recess.]

Chairman TALENT. Let me reconvene the hearing, if I can have order in the room please.

We will reconvene the hearing. I am sorry that I missed Dr. Singer's testimony. I have a markup going on in another Committee and will attempt not to leave in the future.

I do want to take a moment to assure all the witnesses that they will each have full opportunity to respond to what other witnesses are saying. I imagine the questions and answers will bring out all these points.

I will now recognize Dr. John R. Christy, Associate Professor of Atmospheric Science at the University of Alabama at Huntsville.

Dr. Christy, I am very grateful to you for coming all this way to testify, and you may proceed.

STATEMENT OF JOHN R. CHRISTY ASSOCIATE PROFESSOR OF ATMOSPHERIC SCIENCE AT THE UNIVERSITY OF ALABAMA, HUNTSVILLE

Mr. CHRISTY. I would like to move to the projector, if I could.

Chairman TALENT. Sure.

Mr. CHRISTY. Thank you, Mr. Chairman, and Members on Small Business, for inviting me to provide some information on the global climate.

I am John Christy, Associate Professor of Atmospheric Science at the University of Alabama, Huntsville, and a key contributor of the IPCC, which I thought was a credible report.

Please note that I will be referring to figures which you should have attached to your copies of my remarks.

Twenty years is nothing in terms of the Earth's climate, but 20 years is everything in terms of the science of climate, as shown in figure 1.

In the late 1970's, as you see the surface temperature record there, the climate stories were filled with potential disasters of a coming ice age. I lived in South Dakota during the coldest winter of that decade, and we can only hope those ice age forecasts of the seventies will never be realized.

Ten years ago, the potential catastrophe of human-induced global warming began receiving a lot of attention, thanks in part to a couple of warm, but not record breaking, summers in the eastern United States. The predictions were horrifying, rapid temperature rises, coastal flooding, massive hurricanes and so on. Twenty years have now passed since the 1970's and we are now almost exclusively discussing global warming, brought to the forefront 10 years ago.

Twenty years makes all the difference in the science of climate. When global warming became a popular issue 10 years ago, I and others were concerned with the lack of proper data to describe the Earth system and a lack of proper perspective in which to judge the quasi-extreme events occurring at that time, the events so often connected with human-induced climate change. In 1989, NASA scientist Dr. Roy Spencer and I set out to create a satellite data set of truly global observations of atmospheric temperature. These data would not be plagued as the traditional surface data are by changing locations or dependence on transient shipping or lack of coverage for large areas. We did not know what these data would show, whether warming or cooling. Our goal was precision, to provide scientists with excellent, truly global temperature data.

An added incentive for us was that the two layers that we could monitor and observe were two that climate models indicated would show the most rapid responses to climate change: The lower troposphere, the surface at 20,000 feet, which should warm faster than the surface, and the stratosphere, which would cool. In the data presented here, all known artificial effects, such as orbit decay and orbit drift, have been removed.

The top part of the figure there shows the tropospheric temperature since 1979, and it has little trend, either up or down, while the stratosphere, the lower quantity, shows a pretty strong decline, influenced by volcanic and ozone depletion.

Were these data accurate? Balloons also measure the atmospheric temperature at various places around the world, and I and others have compared the satellite data with many other estimates of global temperature, and with this example, done with the United Kingdom Meteorological Office 400-balloon network, you can see the correlation is very high and there is no trend for the period 1979 to 1997. Such independent comparisons demonstrated the satellites were truly providing the precision for which we had hoped. This is that same period in which the science of climate shifted from thinking mostly about global cooling to thinking mostly about global warming.

If we look at the comparison since 1979 between the surface and atmospheric temperatures, the satellite is green, the surface is orange, we see that the trends seem to be going in different directions.

Now, we have only 20 years of satellite data, and that is too short a period to judge long-term climate variations. However, these data tell us that in this layer and during this period when CO2 warming should be most pronounced, we see no trend. We also see something that is not predicted by climate models, a difference in trend between the surface temperatures and the deep atmosphere.

Could the consensus about global warming reached by these same climate models be misrepresentative of the next century if this actual data is not reproduced? It becomes clear that 20 years of truly global data are important.

I am going to change to a different topic here. The recent fixation on extreme events as indicators of climate change is misleading because we know very little about how these events occur and we are able to publicize even marginal extremes to fantastic proportions. Perspective is often lost in the media, though fortunately we have considerable historical information archived at the National Climate Data Center in Asheville, North Carolina.

For example, last month a government-sponsored press conference announced that five States, Maryland, Connecticut, Massachusetts, Vermont and New Jersey, experienced one record high monthly temperature at some point between January and May of this year. Human-induced warming was implicated in that announcement.

But is this significant, since these five States would all fit into the State of Kansas, twice? This sample is not geographically representative.

Please allow me to put a different spin on record high temperatures, using data provided by NCDC, and you will see my point.

Chairman TALENT. What is the NCDC?

Mr. CHRISTY. National Climate Data Center, Asheville, North Carolina. It is where all our climate data is from.

Chairman TALENT. Is that a government operation?

Mr. CHRISTY. Yes, it is.

If you divide the recorded climate history of the entire U.S. into halves in 1940 and look at all-time record high temperatures, you would see that four States had equal highs in both halves. Of the remaining 46 States, 35 recorded their all-time record high temperatures before 1940, and only 11 have since 1940. Or, take the Midwest, this is brand new, newly digitized information, where from the early years now provides a better look at extremes.

For the 10 hottest 3-day periods, these are killer heat waves, in the Midwest since 1896, only two occurred after 1940, and they placed seventh and eighth. Do these results prove cooling is going on? No. They prove that the spin one places on extreme events can be very misleading.

The recent attention given to the heat in the South is another example. A wire service story declared June 1998 the hottest in the history of Huntsville, Alabama. I was surprised that our Office of State Climatologist had not been asked about the record, which I new to be misleading. So I did some checking and found at least 6 other years between 1914 and 1953 in which June was hotter than 1998. I sent the information to the local paper, and they published it. The problem was that a local weather service forecasting office which reporters typically call did not have access to records before 1958 and could not have known about the earlier, hotter years.

Again, the Florida fires and drought were highlighted every evening on the news in June. Was this drought remarkable? The answer from climate records at NCDC is no, absolutely not.

In figure 5, I show the percentage of months in which north central Florida experienced drought conditions decade by decade. In this century, you can see that only one other decade, 1940's, were there fewer drought months than in the 1990's.

Now, yes, it was hot in the South, but the heat wave was due to a weather pattern that placed warm air in the South and cool air in the West. To a climatologist, it was equally remarkable that towns in the High Plains recorded their latest snowfall ever in June 1998, or that Fresno, California, my hometown, for the first time in its history failed to reach 100 degrees Fahrenheit by the middle of July. In fact, and this may seem remarkable to you, in fact, for the Nation as a whole, the temperature in June 1998 was below normal.

Regarding Texas in July, let me say that Dallas is due. North Central Texas Climate Division 8 of the past 10 years have seen colder than average July temperatures. In the past 30 years, 20 have been colder than average. A climatologist sees nothing unusual about Dallas being warmer than average this year. Dallas is due.

To see the real picture of extreme weather variations given by droughts and floods from 1895 to the present, here is figure 6. In this chart, again kindly provided by NCDC, is the percentage of area in the lower 48 States which in any given month experienced extreme conditions, dry or wet. The results show there is no significant long-term trend in either condition. In fact, we have not had a widespread drought in several years, and looking at the chart, we should expect one in the near future, based solely on statistical probability and not on human inducements.

My point is this: Arriving at conclusions about human-induced climate change based on selective extremes is faulty and misleading. In the United States, the evidence presented here suggests human-induced climate change is not yet discernible.

The satellite and balloon data show that significant global warming is not now occurring, though the simple theory and laboratory experiments of greenhouse warming are indisputable. The detection of human effects on the Earth's climate has not been convincingly proven because the variations we have observed to date are not entirely outside of the natural variations of the system. Warming of the planet has occurred on the order of 10 degrees in 40 years in periods of the Holocene here. So I think John has said earlier——

Chairman TALENT. Explain that just a second. You just said the Holocene?

Mr. CHRISTY. In the past 12,000 years, there are records of temperature rises in a period of 40 years of on the order of 5 to 10 degrees Celsius.

Chairman TALENT. When was the Holocene?

Mr. CHRISTY. The last 12,000 years.

Chairman TALENT. You said Holocene. I thought you meant that phenomenon where you can project an FBI picture. I guess that is a holograph, isn't it?

Mr. CHRISTY. A Holocene is a period of climate history.

Chairman TALENT. That is the last 12,000 years.

Mr. CHRISTY. Sorry.

Chairman TALENT. No, that is OK. It is my fault.

Mr. CHRISTY. Another reason why we cannot discern this, or climate models have a little bit of trouble with this, is the Earth's system is very complex and probably has ways to expel heat energy which are not properly accounted for in climate models.

The temperatures in the stratosphere suggest something is going on, but separating the massive effects of volcanoes and ozone depletion, which are effects unrelated to fossil fuel emissions, is not easily done.

I believe some minor warming due to increasing greenhouse gases will occur, though untangling that component from the natural variability would be extremely difficult. Even drastic actions at this point will have a minuscule effect on whatever the present course of the climate is.

We do know that a CO2-enriched atmosphere is beneficial to our carbon-based ecosystems. However, at this point we simply do not know whether the CO2-enriched climate of the next century will be made more difficult or more benign or even noticeably different for our country or for the world. The research is really just beginning.

I will close with a personal observation. Some activists cultivate the notion that they are taking the high moral ground by putting global warming on the same "save the Earth platform" as preserving wildlife, saving rain forests or cleaning the Earth's rivers and oceans.

Serving as a missionary in Africa in the early 1970's gives me a different perspective on proposals to combat global warming. The artificial rise in fuel and energy costs created by the Arab oil embargo was a crushing blow to many, but in particular to the world's poorest peoples. Many people in Africa suffered or died when gasoline prices soared beyond their ability to pay, halting transportation and other energy-related activities that preserve both life and property. I am well aware of this personally because people would come to my door and bang on it in the middle of the night to try to get an injured or very ill person to the clinic, and many of them were unable to because there just wasn't any gasoline.

In addition, many of the life saving activities supported by the benevolence of wealthy countries, such as ours, were curtailed due to the rising costs and slumping economies back home.

Disrupting the lives of those whose existence is too often literally hanging by a thread causes the kind of suffering that the average policymaker or activist never sees. I have seen it. It is real and it is devastating.

That is why my own view, based on personal experience as both an atmospheric scientist and a missionary, leads me to a different high moral ground, which is this: When it comes to global warming, save the Earth, but not before you reduce the scientific uncertainties and not by sacrificing the poor.

Thank you.

[Mr. Christy's statement may be found in the appendix.]

Chairman TALENT. Thank you, Dr. Christy. Our last witness is Dr. Marlo Lewis, Jr., the Vice President for Policy and Coalitions of the Competitive Enterprise institute in Washington, DC.

Dr. Lewis.

STATEMENT OF MARLO LEWIS, JR. VICE PRESIDENT FOR POLICY AND COALITIONS OF THE COMPETITIVE ENTERPRISE INSTITUTE IN WASHINGTON, DC

Mr. LEWIS. Thank you, Mr. Chairman and Members of the Committee, for this opportunity to present testimony on the global warming debate. What I would like to do in the next several minutes is address what I take to be the Clinton administration's strongest argument in favor of a climate treaty. This is the argument that, if we just do it smart, Kyoto will provide low cost planet insurance for our children and future generations.

In candid moments, Kyoto partisans may admit that the science of climate change is not really settled. But they don't see this as a great liability. Their trump card argument in the global warming debate is not any testable scientific hypothesis, but something called the precautionary principle. This is the proposition that lack of scientific certainty should not become an excuse for inaction where there are threats of serious or irreversible harm to health, safety, or biodiversity.

The precautionary case for Kyoto goes as follows: The Earth may be warming, industrial activity may be the cause, and the effects of such warming may be catastrophic. Mankind has been running a gigantic uncontrolled experiment on the climate system. Since the experiment is potentially life threatening, we should start applying controls now. The alternative is to throw caution to the winds and gamble with the only planet we have. Kyoto or some similar treaty is the only responsible option.

This argument is rhetorically powerful because it sounds like the familiar maxim of common sense, err on the side of caution. However, as will soon become clear, we can with equal legitimacy invoke precautionary considerations to oppose the Kyoto Protocol.

My testimony has two parts. First, I will show that the precautionary principle supplies no guidance for choosing between policy alternatives. Then I will offer specific reasons why Kyoto is not an assurance policy.

The fatal flaw in the precautionary case for Kyoto, as in environmental advocacy generally, is its complete one-sidedness. Environmentalists demand assurances of no harm only with respect to actions that government might regulate, never with respect to government regulation itself. But government intervention frequently boomerangs, creating the very risks that precautionists deem intolerable.

For example, banning DDT has contributed to the revival of malaria epidemics in the Third World. Frank Cross at the University of Texas at Austin notes that overregulation can kill just by misdirecting resources. Regulatory schemes that divert attention, effort and money from major threats to minor risks can make us less safe. For example, the millions of dollars that local governments waste on gold-plated Superfund cleanups cannot be used to improve police and fire protection.

More importantly, for individuals as well as nations, wealthier is healthier and richer is safer. There is an obvious connection between livelihoods, living standards, and lives. Wealth is the single most important factor affecting health and longevity. Even in comparatively affluent countries like the United States, studies indicate that every $5 to $10 million dollar drop in economic output translates into one statistical death.

So here is a precautionary case against the Kyoto Protocol. Stabilizing greenhouse gases at levels low enough to cool the planet may require drastic reductions in energy use. An energy constrained world may be a poorer world. And a poorer world may have more starving people.

True, I have no scientific proof that Kyoto would condemn millions of people to poverty, starvation and misery, but the precautionary principle says we should not let the absence of scientific certainty become an excuse for inaction. It also says we should not permit far-reaching innovations until they are proved to be safe. No one has demonstrated that Kyoto will not have harmful consequences. Therefore, we should oppose it.

Kyoto's partisans admonish us not to gamble with the only planet we have, yet they are perfectly willing to gamble with the only economy we have. They cannot logically have it both ways.

I do not mean to suggest that the precautionary principle would be beneficial were it applied evenhandedly to economic and regulatory activity alike. Inflating "Safety First!" from a mere rule of thumb into a categorical imperative, an absolute overriding duty, is a recipe for paralysis and stagnation, perhaps the riskiest condition of all. My point, rather, is that the precautionary principle is neither a scientific nor an ethical precept, but a rhetorical weapon. Its purpose is to exaggerate the risks associated with economic activity and conceal the risks arising from the exercise of political power.

The administration claims that Kyoto is an insurance policy, and insurance by definition is supposed to make us safer. But is Kyoto the real thing, or is it as phony as the precautionary principle?

Whether or not a proposed insurance policymakers sense depends chiefly on three things: The probability and size of the potential losses to be insured against, the cost of the premiums, and the extent of the coverage. Let's examine Kyoto under those three aspects.

How likely is catastrophic warming?

The computer models on which the Rio Treaty was based overestimate by 200 to 300 percent the amount of warming recorded over the past century. If the models can't hind-cast the past, why should we suppose they can accurately forecast the future? When the planet was warmer than today, for example, in the Medieval Warming of 900 AD to 1200 AD, mankind flourished. When the planet was cooler, for example, the Little Ice Age of 1300 AD to 1850 AD, mankind suffered.

I conclude from this and other facts which have been presented today that the probability of catastrophic warming is low.

How much would Kyoto insurance premiums cost? The administration says not more than one-tenth of 1 percent of GDP. But this estimate is based on unrealistic assumptions, such as the operation of a full-blown emissions trading system that includes key developing countries. The estimate also overlooks the enormous regulatory burden that will hit over 1 million small businesses if EPA makes good on its threat to classify CO2 as a hazardous pollutant. As energy analyst Mark Mills shows in the latest World Climate Report, wind and solar power would have to increase their output 37,000 times just to meet 30 percent of the expected growth in electricity demand over the next 25 years. Clearly Kyoto's premium payments will be higher than the administration's estimates, perhaps staggeringly so.

How much protection then would Kyoto insurance provide?

We are really asking two questions here. First, how much would Kyoto reduce the amount of global warming, assuming for a moment that the climate models are right? As Dr. Michaels has shown, Kyoto under the most generous assumptions would cool the planet by less than one-tenth of a degree Celsius.

Second, would Kyoto help make us whole after global warming has occurred. Here is where the insurance analogy completely breaks down. Kyoto can do nothing to put us back on our feet once we have suffered damages from catastrophic climate change. Quite the contrary, whatever resources we apply under Kyoto to prevent climate change, we cannot use to adapt to climate change if and when it occurs. Kyoto is all premium and no coverage.

This last point leads to a question that should be at the heart of the global warming debate: Which type of social insurance policy is likely to deliver the most protection? Should we try to prevent global change by rationing energy and further politicizing economic and technological development, or should we try to increase mankind's ability to adapt to change by reducing political barriers to enterprise, innovation and invention?

Under the latter option, let's call it a resiliency strategy, policymakers would work to eliminate the high taxes and burdensome regulations that impede economic growth and technological process. Wealthier, more technologically advanced societies, are better able to anticipate, withstand and recover from extreme weather events and other natural disasters.

A resiliency strategy is clearly the superior option, given the uncertainty surrounding the global warming hypothesis. Many possible catastrophes may befall us in the next century, the outbreak of new viral plagues, increased tectonic plate activity, an errant meteor, nuclear terrorism, chemical or biological warfare, even, as Dr. Singer pointed out, global cooling.

Making societies freer and wealthier is inherently desirable, and would better prepare us to deal with whatever shocks and surprises the future may hold. In sharp contrast to Kyoto, this resiliency approach is all coverage and no premium.

The Kyoto Protocol would restrict our freedom, diminish and possibly crush our prosperity, and mobilize tremendous resources to fend off a threat that may prove to be nonexistent or trivial compared to the age old scourges of poverty, hunger and disease. America does not need and cannot afford that kind of insurance.

[Mr. Lewis' statement may be found in the appendix.]

Chairman TALENT. I thank the gentleman for his testimony. We will go to questions now. I believe I will defer my questions, since I know that other Members have to leave. If other Members leave, it is going to be pretty lonely up here.

I will first recognize the distinguished gentlewoman from New York, the ranking Member, Ms. Velazquez.

Ms. VELAZQUEZ. Thank you, Mr. Chairman. I would like to direct my first question to Dr. Lashof and Dr. Watson. It would appear that most other nations in the world have accepted the scientific argument which maintains that human-induced global warming is in fact happening. Do you have a sense of the rough percentages of scientists with expertise in this area, both worldwide and in the United States, who support your views?

Mr. WATSON. I don't know there is a real answer to that. I do believe that the IPCC is a process that reaches out to the very best scientists, wherever they are in the world, whatever type of organization they work for, universities, government labs or industry, and we try to very carefully assess what is known and what is not known, which is why I do believe that the view of the IPCC was endorsed by literally all governments as the basis for Kyoto. It (IPCC) does not come up with a single number. It (IPCC) has a range, and that encompasses some of this uncertainty.

So I would not give a percentage. It is very hard to. I would personally say the very large majority of scientists.

Mr. LASHOF. I agree with Dr. Watson's assessment. Surveys of scientists can be extremely misleading. Dr. Singer referred to one petition that in fact was sent out to, I don't know how many hundreds of thousands of people. Some people signed it, but there is no screening, no assessment. What we need is a rational process to assess the data. Again, that is why the Reagan administration, was originally behind setting up the IPCC, it was supported by the Bush administration, and supported not only by our government through different administrations but by governments around the world. It is, I believe, the appropriate way to assess global warming science. Nobody that has criticized it has suggested a better alternative.

Ms. VELAZQUEZ. Is opposition greater in the United States than in other industrialized nations?

Mr. LASHOF. I think that there is more of a political debate over the science in the United States than in other countries. I think that is fueled clearly by special interests that have amplified the views of a very small handful of skeptics, such as those represented on this panel, well beyond their representation within the scientific community. That phenomenon is not completely unique to the United States, but it certainly is happening more in the United States than in other countries.

Ms. VELAZQUEZ. In your view, can emissions of greenhouse gases be reduced without significant costs to the economies of developed and undeveloped nations?

Mr. LASHOF. I believe they can. I did not incorporate economic analysis in my testimony here today, but I would note that the Union of Concerned Scientists released a report recently assessing the economic costs or potential benefits of achieving the Kyoto Protocol. I think that in fact the costs are very modest, at most on the order of plus or minus a tenth of a percent of GDP.

This is a copy of the Union of Concerned Scientists' report which reviews some work we were also involved with, called "A Small Price to Pay," which I think documents that from a large literature of studies that look carefully at the technologies that are available to reduce greenhouse gas emissions.

Ms. VELAZQUEZ. Thank you. Dr. Michaels, it is my understanding that you are the editor of the World Climate Report. Is this true?

Mr. MICHAELS. That is correct.

Ms. VELAZQUEZ. Let me read to you a section from the September 1st, 1997, edition from an article entitled "Some Like It Hotter."

"Heat stress does increase mortality, but it affects typically only the old and infirm whose lives may be shortened by a few days or perhaps a week."

Would you please comment on this?

Mr. MICHAELS. That is a well-known phenomenon. It has been studied by Larry Colkstein from the University of Delaware under EPA contract. It is called mortality harvesting; that when you have an environmental stress, what happens is the death rate goes up, and then after the stress declines, the death rate drops below where it was.

I want to talk to you. You have brought up an incredibly important subject. The newspapers, this week I think, last Friday, most of them carried a story that heat-related death was the largest single cause of natural hazard mortality. Do you remember those stories? It alluded to a heat wave in 1901 that killed 9,700 people.

Now, we have had many heat waves of the magnitude of 1901, some quite recent. The one we are having now is not of that order. In 1995 it was. The total deaths run about 100 to 500.

What we have done in our society, if we extended the population from 1901 to now, I need to make my point, the deaths that you would get from that heat wave would have been between 50 and 100,000. Yet we see 50 to 100. Our society has buffered itself against the vagaries of nature with the same technology that has enhanced the greenhouse effect.

Ms. VELAZQUEZ. I want to go back to your article. Can you explain why the old and infirm and people in developing countries should be exposed to increased health risks simply so industry does not have stricter emission standards?

Mr. MICHAELS. Again, I direct you to the work of Dr. Colkstein, that whenever there is heat-related stress, there is a change in the mortality that shows up as a deficit in death immediately after the stress declines. These deficit periods are on the order of 2 to 3 weeks.

Mr. LASHOF. Pat, I have to respond to that. You are right, Kalkstein documented that, but he also documented taking that into account, there is significant health increases in mortality that is not associated with that phenomenon. So if you want to cite Kalkstein, cite everything he says about the subject, not just one part like you did with Hansen, where you take one sentence out of the paper and leave the rest of it out.

Mr. MICHAELS. I would caution you the weather-related death rate in the winter is greater than the summer, the mortality rate. So if you decrease the severity of the winter, which is what I believe you do by enhancing the greenhouse effect, and so does basic physics, you have affected the mortality in a positive sense.

Ms. VELAZQUEZ. Dr. Watson, would you care to react to his statement?

Mr. WATSON. I think what Pat is trying to say is we as a society have buffered ourselves against the very large threat of heat stroke mortality through the use of energy systems, giving us air conditioning. The challenge is getting the energy in a much cleaner way, more efficient systems and cleaner production of energy.

So the point is we should not be threatening the old, infirm and sometimes the very young and people in developing countries. The challenge isn't air conditioning or not air conditioning. The challenge is how do we get air conditioning in a way that doesn't cause global warming.

Ms. VELAZQUEZ. Dr. Michaels, who sponsors and funds publication of this report?

Mr. MICHAELS. That is produced by New Hope Environmental Service, which is a consulting company.

Ms. VELAZQUEZ. Pardon?

Mr. MICHAELS. New Hope Environmental Service, a consulting company, which receives a contract from the Greening Earth Society to pay for this report.

Ms. VELAZQUEZ. I thought it was funded by the Western Fuels Association.

Mr. MICHAELS. No, Greening Earth Society now.

Ms. VELAZQUEZ. Dr. Singer, peer review is one of the most important aspects of ensuring that scientific claims are valid. Would you briefly explain peer review to this panel?

Mr. SINGER. Peer review is a device used by editors of scientific journals to help them in making a decision whether to publish a piece of original research. It means that the editor will send this article, paper, this scientific research paper, to referees who remain anonymous and who give him his comments.

The final decision is up to the editor. The editor decides who the referees are and whether in the final analysis to publish the paper.

Ms. VELAZQUEZ. Dr. Singer, have the findings and conclusions in your testimony been published in scientific peer review journals?

Mr. SINGER. My findings and conclusions are based on published peer reviewed articles, yes.

Ms. VELAZQUEZ. Can you mention traditional scientific journals?

Mr. SINGER. Yes. I can show you the references that I refer to. These include all of the major scientific journals, the same journals, incidentally, used in the IPCC report. They include the Journal of Science in the United States, the Journal of Nature in Great Britain, and specific specialized journals dealing with climate issues and geophysical issues.

Ms. VELAZQUEZ. Dr. Watson?

Mr. WATSON. Dr. Singer beautifully avoided your question. What you asked has his analysis been peer reviewed, and I would argue his analysis has not undergone the same peer review as the articles in Science or IPCC document. So he purposely avoided the question.

Mr. SINGER. My analysis, by the way, is not an original piece of scientific work. I am very careful to state this. This is based on published peer reviewed work that appears in the literature.

Ms. VELAZQUEZ. So your conclusion has not been submitted to peer review?

Mr. SINGER. My conclusion doesn't have to be peer reviewed. It is not a piece of scientific research.

Ms. VELAZQUEZ. You explained when I asked you what is peer review, you explained it was a very important aspect.

Mr. SINGER. Yes, for original work.

Mr. LASHOF. I would just point out that review articles, which are not original pieces of research, but review the literature, are also traditionally subject to peer review when they are published in scientific literature.

Mr. MICHAELS. I would also point out that books published are not peer reviewed, and Mr. Singer published a book. Any book that is a summary of peer reviewed research is not going to undergo the same process. All he is doing is restating the arguments that have already been peer reviewed.

Ms. VELAZQUEZ. We are discussing here scientific data and conclusions and findings. We are not discussing here publication of books.

Mr. Chairman, Dr. Singer mentioned 15,000 signatories of the Oregon Institute of Science and Medicine Global Warming Petition Project. For the record, I would like to note that the petition uses an op-ed article published in the Wall Street Journal as its basis. Also I would like the signatories' names entered into the record, and I have them here. I would like to read a couple of familiar names: Dr. Honeycutt, Dr. Pierce, and Dr. Potter. It seems to me that these are from the cast of MASH. Joseph McCarthy, whom I believe died even before the petition was circulated.

[The information may be found in the appendix.]

Ms. VELAZQUEZ. My next question is to Dr. Christy. You claim that satellite data shows that the Earth is actually cooling. Is this your position?

Mr. CHRISTY. Through June 1998, the trend is plus .04 degrees C per decade.

Ms. VELAZQUEZ. Isn't it true that only one satellite, the MSU–2R, shows this cooling trend?

Mr. CHRISTY. No, there are nine satellites. There are only two at a given time.

Ms. VELAZQUEZ. According to an article published in Nature Magazine from March 1997, this cooling trend coincided with satellite replacements. If adjustments are made, the article concludes that MSU temperatures actually show a slight warming of the Earth. Would you care to comment on this?

Mr. CHRISTY. Yes. I refuted that in the Journal of Nature in September 1997 and a Journal Climate article that will be out next month, and it shows clearly based on independent evidence that that analysis was incorrect.

Ms. VELAZQUEZ. Dr. Christy, if there is a cooling process going on, could you explain why the National Oceanic and Atmospheric Administration announced that 1997 was the warmest year on record?

Mr. CHRISTY. Their analysis is based upon several thousand reports from surface measurements. It was not in the atmosphere. Whether you use balloons or satellites, 1997 was a very average year globally. I believe both of those, because their precision is documented and they are, or at least the satellites are globally measuring the thing.

Ms. VELAZQUEZ. Dr. Michaels, Dr. Singer and Dr. Lewis, could you tell the panel about your sources of funding?

Mr. LEWIS. Yes. The Competitive Enterprise institute receives funds from individuals, corporations and foundations. We get no government support.

Ms. VELAZQUEZ. Dr. Singer?

Mr. SINGER. The same.

Ms. VELAZQUEZ. You have responded before.

Mr. MICHAELS. No, I did not. No. All my academic research is funded by taxpayer-supported entities, including NOAA, the Department of Energy, Commonwealth of Virginia, National Weather Service, et cetera. Thank you.

Ms. VELAZQUEZ. It is also true you have received money from Texaco, German Coal Mining Association, Western Fuels Association, the American Petroleum Institute and other fuels companies?

Mr. MICHAELS. Who are you asking?

Ms. VELAZQUEZ. Dr. Michaels, Dr. Singer, and Dr. Lewis?

Mr. MICHAELS. What you have on the record, my funding history, the chairman has it. Again, you will see that the academic research that I do, which is published in the peer review literature, just like everybody else, is supported solely by public sources. I am sorry.

Mr. LEWIS. It is CEI's policy not to discuss the identity of its donors in public, but I will just make the observation that if you are a nonprofit organization, you can basically get money from only two places. It is either business or someone retired and wealthy who has been in business or the government; i.e., the taxpayer.

Ms. VELAZQUEZ. My question is if you receive money from Western Fuels Association, Texaco, that many of these groups will be strongly affected by a required lowering of emissions. Doesn't this seem to be a conflict of interest?

Mr. LEWIS. Not at all. At the Competitive Enterprise Institute we consider ourselves a "battered business bureau." But more precisely, were a pro-market organization. There are times, unfortunately not enough times, when corporate America actually favors free market policies. On those occasions, we sometimes earn their support. We think they should be giving us even more money than they do, because we think that if we really did move to a freer world, it would be better for everyone, including General Motors. But sometimes they take a very short-term view and are more concerned about a piece of corporate pork, which we oppose, than they are about more fundamental reform.

Ms. VELAZQUEZ. Mr. Chairman, I have finished with my questions. My question was basically is it a conflict of interest, to try to show a conflict of interest. If you are going to get money from the same sources and you are going to produce some conclusions that are going to benefit in terms of any regulations that will have an impact, a negative impact on those corporations, it will be an ethical question.

Mr. LEWIS. Yes. We have a consistent principled approach to public policy. We have taken the approach that I have been developing here on global warming before any corporations were interested at all in the work we were doing. The first article we published on this subject was 1991. That was before we were getting any support that was in any way related to our global warming work, and as I said, at times businesses will see that their own bottom line considerations correspond to the principled policies that we promote. We also make it a policy never to do any contract research. In other words, a corporation cannot pay us to do a study. We have an agenda, and if they like it, they can support it.

Ms. VELAZQUEZ. Sure. Mr. Chairman, I would like to enter into the record this information that has been provided when they have to file for tax documents. The Competitive Enterprise Institute is a nonprofit industry think-tank that receives a great deal of its funding from the fossil fuel-related companies and groups, American Petroleum, Amoco, ARCO, Ford Motor Company, General Motors, Texaco among others.

Chairman TALENT. Without objection, it will be entered into the record. The source of that information was the tax documents they filed?

Ms. VELAZQUEZ. Yes.

[The information may be found in the appendix.]

Mr. SINGER. I would like to say that none of the entities mentioned by the lady has been supporting us. I would like to correct my statement, however, to mention that we have received a contract from the Commonwealth of Virginia, so it is taxpayer supported, to study what to do about improving air quality and what to do about traffic congestion in Northern Virginia. Otherwise I have no further comments.

Chairman TALENT. I recognize the gentlewoman from Missouri.

Mrs. EMERSON. Thank you, Mr. Chairman. Let me just follow on that last question and take a different approach.

Obviously there are some scientists who are skeptical of the presence of global warming. We saw that today, especially those who criticize the protocol as often being dominated by corporations.

Dr. Christy, you were part of the IPCC. You are a signatory to the IPCC document, is that correct? What can you tell me about the qualifications of the 2,600 scientists who signed the IPCC?

Mr. CHRISTY. There was a broad range, there was a broad range of people involved. Many were just government appointees and had little background in this science at all. In our chapter if we had probably 150 people involved in some way, often that was simply a 2 or 3 sentence refereed or reviewed sentence, and that got their name into the list. But really the work was done by on the order of 10 people in our chapter. There were 11 chapters in the working group 1. Bob Watson would know more.

Mrs. EMERSON. Perhaps I should ask Dr. Watson that question too. I was interested in your perspective.

Dr. Watson, tell me about the background, generally speaking, of those 2,600 so-called scientists.

Mr. WATSON. The majority were from universities and government organizations, a small number from environmental NGO's and business. They were sponsored, or nominated rather, either by governments, intellectual think tanks, organizations such as International Council of Scientific Unions, and they were selected by the IPCC bureaus. So there is a very wide range.

My personal view is we need more people who actually come from business and industry and even are paid for by business and industry to be a part of IPCC to see the full range of views. So, to be quite candid with you, source of funding is not my concern. My concern is to make sure that we have a broad range of representatives who are credible in the scientific and expert community.

Mrs. EMERSON. So of those 2,600 people, how many were climatologists?

Mr. WATSON. There is a complete range. You see, IPCC looks at the whole range of issues from: Is the climate changing, to what are the economic costs. So the climatologists, there may be 20 to 50. There are some oceanographers and mathematical scientists; There are also economists and social scientists.

So there is, in essence, in the Second Assessment I would guess 50 chapters. Each of those chapters needs a totally different mix of skills. So all 2,600 are clearly not climatologists.

Mr. SINGER. Dr. Watson I think is evading the question. I have made a careful study of this and have published an article in the *Wall Street Journal*. There are no 2,600 scientists. That is complete fiction.

Another piece of fiction is that there is a consensus amongst these IPCC participants. There was never any kind of polling or any kind of opinion poll done on it.

When you analyze the number of names mentioned in the three IPCC volumes that were published in 1996, the number is of the order of 2,000. When you eliminate the duplications among the names, because they appear in different volumes, the number falls well below 2,000. Most of these 2,000 are not scientists, physical scientists or climatologists. There are social scientists, political scientists, economists, government people, policy analysts. There is even a person, as a reviewer, who is the public relations chief for the American Petroleum Institute listed as one of the participants of the IPCC.

So we have a wide range of people. The actual climatological people who one could identify amount to probably not more than 100 to 200.

Mrs. EMERSON. The only reason I wanted to ask that is because Carol Browner has on many occasions said how this 2,600 list of scientific experts in the field, but I have noticed that there is an OB–GYN on the list and a psychologist on the list and an acupuncturist on the list, none of which I have any problems with.

50-206 98 - 2

Dr. Michaels.

Mr. MICHAELS. I think you may be interchanging two separate lists, and I think Ms. Browner may be doing the same.

There was a petition sent to various scientific organizations, including the American Geophysical Union, the Ecological Society of America. That has nothing to do with the Intergovernmental Panel on Climate Change. It was done by a group called Ozone Action, which is a lobbying organization, and they got 2,600 signatures, which is a remarkably small number given the large number of people that belong to these organizations. I hesitate to give an exact number, but I think it was certainly less than one out of ten.

Mrs. EMERSON. That is a very good point you make and I really appreciate your bringing that to my attention because perhaps you are right. She has also tried to confuse the issue.

Let me just move on to another question here. This Committee has identified at least $6.3 billion that the administration has requested to implement or would like to spend in the next fiscal year to begin implementing the Kyoto Protocol. Yet, obviously, as I said earlier, you all disagree as to whether or not we have a problem with global warming.

Can any of you all give me any indication at all of how much it would cost and how long would it take to really determine whether or not the Earth's temperatures are rising, whether they are falling or whether they are staying the same? Is it possible to get our arms around this, and how long would it take and how much would it cost?

Mr. SINGER. These problems cannot be settled by just throwing money at them. They require observations of the climate. You cannot hurry these things. They take time.

For example, Dr. Christy pointed out that currently the climate is not warming. In fact, it has been cooling, as seen by satellite data. Throwing money at this does not affect the speed at which you can gather these data. They will be coming in year by year as the climate develops. Incidentally, the climate is going to change no matter what. It will always either get colder or warmer. It can never be stabilized.

Mrs. EMERSON. Dr. Watson.

Mr. WATSON. The basic point is, over the last 100 years, the temperature has warmed. Over the last 20 years, there is clearly a discrepancy between the ground-based observations that still show a small warming and the satellite observation that basically show no warming and possibly a slight decrease, possibly a very small increase.

So there are two questions. First, is the Earth's climate changing? But the second one, which is equally and even more important is, why is it changing? So one has to look not only at the data to see if there is a change, and I would argue we have to look at the long-term change, not the short-term 10 or 20 years because of natural fluctuations in the climate. But the key issue is establishing cause and effect.

That is indeed a long process if indeed you want absolutely no ambiguity on this issue. The trouble is the time scale of the climate system is so long, because of the lifetime for CO2, the challenge for policymakers such as yourself is that you are going to have to make some decisions without perfect knowledge.

Mrs. EMERSON. Dr. Singer.

Mr. SINGER. Again there may be some confusion here.

The Federal Government does have a very active and broad research program now which is of the order of $2 billion a year. We are not talking about that. That is the program that is supposed to determine what is causing climate change, among other things. So that deals with the science, or should deal with the science. The $6.3 billion you are speaking of——

Mrs. EMERSON. I realize that is totally separate.

Mr. SINGER [continuing]. Is unfortunately not being spent for science at all. It includes corporate welfare programs. It includes subsidies. It includes many other things not having to do with science.

Chairman TALENT. Would the gentlelady yield?

$2 billion you are spending a year, right? Are you getting any of that money?

Mr. SINGER. I get none of this money.

Chairman TALENT. Somebody is getting that $2 billion, aren't they? If we were to conclude that global warming is not a problem, would we be spending the $2 billion?

Mr. SINGER. Probably not. I might point out that this is more than is being given to the National Cancer Institute, which has a smaller budget. So this tells you something about perverted priorities of the Government.

Chairman TALENT. I will wait for my questions. I yield back.

Mrs. EMERSON. Dr. Lashof, you had your hand up.

Mr. LASHOF. I would like to address the $6.3 billion proposed. The Clinton administration's Climate Technology Initiative is what it is called.

You characterize it as proposed spending to implement Kyoto and I think that is incorrect. As far as I know, the administration has not proposed any funds to implement Kyoto. The budget for fiscal year 99, That is being debated now, obviously was prepared before the Kyoto Protocol was in fact negotiated, although it was released afterwards.

The point I want to make is, if you look at what is proposed there, it is primarily for energy efficiency, renewable energy, both technology research and development, much of which is going to scientists, as well as tax incentives, it is fully consistent; indeed, one could argue the kind of program that is required by the 1992 Framework Convention on Climate Change, which is the law of the land. It was ratified by the Senate in 1992.

Mrs. EMERSON. But my question really had to do with if in fact you all as scientists could say, yes, we can in fact determine over this length of time that there is a problem or there is not a problem and then how much it would cost.

Dr. Lewis.

Mr. LEWIS. Yes, I would like to speak just briefly to the question of priorities. We heard comments here today that one of the perils potentially of climate change is an increase in disease vectors or the spread of malaria and dengue fever and so on.

Just yesterday, there was a fascinating presentation given by Dr. Paul Reiter of the Centers for Disease Control, who rejects entirely the thesis that climate change is in any way important to the public health of the United States and not even the developing countries with respect to diseases.

He pointed out to me, though, that right now the U.S. Government is only spending $200,000 a year to try to develop a vaccine for dengue fever, which is really having an upsurge in places like Puerto Rico, where he is stationed. I think we just have to ask ourselves, how many $200,000 grants is the Federal Government handing out to nonprofit organizations to study climate change, or how many times does $200,000 go into $6.3 billion?

I mean, if we are really concerned about the health effects of climate change, why aren't we even more concerned about the suffering and dying of people right now and why don't we divert some of that $6.3 billion into developing a vaccine for dengue fever if dengue fever and the threat of it is one of the rationales given for the Kyoto Protocol?

It seems to me that some of the priorities here are completely twisted.

Mrs. EMERSON. You make a good point. If you take a look at the Environmental Protection Agency's budget and that particularly which has to do with studying global climate and greenhouse gas impacts, there are many, many grants to many different organizations but they are mostly advocacy groups who have an unbalanced position already, but there are several million dollars worth of those grants.

Dr. Watson, you mentioned during answering the last question the increase in greenhouse gases produced in the last century. You talked about that. Can you measure how much of that increase is due to human activity?

Mr. WATSON. Of carbon dioxide, we are almost sure that nearly all of it is due to human activity. One can look at that by, without getting too technical, the so-called isotopic composition of the change in carbon dioxide in the Earth's atmosphere. Probably 70 percent or thereabouts, I have not got an exact number, has probably come from the burning of fossil fuels. That is the combustion of coal, oil, and natural gas. Some of it has come from deforestation, that is cutting down trees, and most recently in the tropics.

Just as one quick aside, and I apologize for this, the U.S. Global Change Research Program is correctly stated to be between $1.8 and $2 billion a year. The National Cancer Institute Budget is $2.54 billion, so they are comparable. I think that is a fair thing to say.

Mrs. EMERSON. So, anyway, you are saying that perhaps most of it is due to human activity. But yet, in your 1995 IPCC, if I can quote from page 5, it says "Our ability to quantify the human influence on global climate is limited because the expected signal is still emerging from the noise of natural variability and because there are uncertainties in key factors, etc."

So, in other words, you think that it does but you are not sure?

Mr. WATSON. They are two different questions, good questions, both of them.

The first question is, has atmospheric carbon dioxide and other greenhouse gases increased? The answer is unambiguously yes. No one would debate that. I believe it is also fair to say most people would agree the carbon dioxide increase is almost exclusively due to human activity. So atmospheric composition is changing.

Second, has the global temperature increased? The answer is also yes. Third, I will get to your last point, is: Can we establish cause and effect between the greenhouse gases increasing and the observed change in the world's temperature?

We have said, basically, there are uncertainties. But the preponderance of scientific evidence supports the view that human activities are involved in the observed climate change. So there are really three steps to the argument.

Mr. MICHAELS. Could I amplify on that for a minute? Because I think Bob is accurate except he left something off, which is, while one might say that the evidence suggests a "discernible" — that is the word they used — human influence on the climate, the real question is why is it so little and because it is so little, what is nature trying to tell us?

Again, this constant error factor of about a third between what was predicted and what was observed remains, and will, and I think tells us a lot about this issue. You do not care whether it warms. Nobody here cares whether we warm the climate. You care whether we warm it a lot; you care whether we warm it in a deleterious fashion.

If the human warming is primarily in the coldest air masses of the winter, and physics would say that is true, I am not so sure that is such a terrible thing.

Mrs. EMERSON. Of the warming that has taken place in this century, would you agree, Dr. Watson, that most of it has happened pre-1940?

Mr. WATSON. If you look at the data you could say probably about half occurred pre-1940, half since. The argument is nothing should have occurred pre-1940. That is not true. There was an increase in carbon dioxide and other gases, so some of the warming would have been expected before 1940.

The theoretical models that try to compare theory and observation have taken into account not only the change in greenhouse gases but the change in sulfate aerosols gases and they actually get a fairly good correspondence, not a factor 3 discrepancy as stated by Pat Michaels, when you do it carefully all the way from 1900 all the way through today. In fact, the projections that were used for Kyoto have taken these models that have had less of a projection than 1990 and there is a fairly good correspondence between the latest models and the observations.

Mr. MICHAELS. I must respond. Bob Watson just made the assertion that when you adjust the climate models for sulfate aerosols that you fit the globe's temperature much better.

Now, there are two hemispheres on this planet, the northern and the southern. Almost all the industry is concentrated in the northern hemisphere. The southern hemisphere has very little industry and, therefore, very little sulfates that would result in a canceling of the greenhouse warming.

If you will take a look at my testimony, please, on page 4, you will see that the southern half of the planet is cooling. There are no sulfates there to cool it. That is prima facie evidence that the sulfate hypothesis is in big trouble. If this discussion goes much further, I will show you that it is in even bigger trouble than that.

Mrs. EMERSON. I just want to ask one more question, because the 1990 IPCC scientific assessment says that the rather rapid changes in global temperature around 1920 to 1940 are likely to have been caused by natural origin.

So what is natural origin?

Mr. WATSON. This is the point that one has to make. For any short period, say 10 or 20 years, there could be human-induced changes for expand the period 1920 through 1940, that would be relatively small because of only a small increase in carbon dioxide.

There are also a natural exchange of energy between the atmosphere and the ocean. There is potential small changes in fluctuation in the solar output, changes in volcanic activity that change aerosol particles in the atmosphere, and also changes in El Nino event, which is a natural climate oscillation.

So it is not at all surprising that over any 10–20 year period you may get a warming or a cooling as much as a tenth to three-tenths degree Centigrade even in the absence of human interaction.

Mr. MICHAELS. Could I respond. I would like to say a couple things. First of all, the warming for 1920 to 1940, which the IPCC said was likely to be natural, in fact is almost exactly equal to the warming that occurred between 1970, in the last 20 years or so. But I want to tell you how much we understand this issue.

We have temperature records taken by weather balloons and, as you know, they show no warming since 1979; and they match up perfectly with the satellite record that John Christy monitors. However, if we take a look at the weather balloon record back to when it begins, and it begins in 1958 globally, there is a warming in it. That means that there had to have been a warming prior to 1979 that was pretty sharp.

In fact, there was. It occurred in 1 year, 1976, 22 years ago. There is a jump in the temperature that took place 22 years ago, in the weather balloon record of a lower atmosphere. All the readings before 1976 are flat. All the readings after 1976 are flat. I do not know of one climate model that says that anthropogenerated warming would take place in 1 year 22 years ago.

Mrs. EMERSON. Well, there is so much confusion here that I am very reluctant to spend $6.3 billion or even more than that unless you all can decide whether or not there is a problem. It is kind of like dairy farmers. You know, you cannot get two dairy farmers who agree on anything even if they live next door to each other.

So I am getting the impression that you all do not agree. So, therefore, I am very reluctant for us to spend a whole lot of money trying to implement something through the back door, which I think is perhaps, even though Dr. Lashof and I might disagree on that, having looked at old 1994 EPA documents that pretty well suggests that this is what they intend to do, I am reluctant to spend more money on something unless we know what the problem is and how to fix it or whether or not there is a problem at all.

Mr. WATSON. Let me make a quick comment that is neither to attack or defend that particular $6.3 billion, except to say, some of that money is to be aimed at research and development of energy-efficient technologies, renewable energy technologies. They make good common sense with or without climate change in the sense that we should not be as dependent as we are on foreign oil.

So if indeed we can invest in technologies for the future, they are a good investment both to hedge ourselves against disruption in foreign oil, but they are also a good investment in case we do need in the future technologies such as these to address the issue of climate change.

So I think one has to look to see what exactly is being proposed on the technology side, and I would argue that an investment in technologies that can make our society more efficiently use energy, and be less dependent on fossil fuels makes sense anyway.

Mrs. EMERSON. Do you think that should be the Government's role and not the private sector's role?

Mr. WATSON. I think there is a role for both Government and for the private sector. In fact, if one looks, I think one has to note that it is getting more difficult for the private sector to invest in long-term technology development when they may or may not be able to appropriate the value of that research.

So I think there is a balance and an ideological question that this Congress has to wrestle with. I believe personally there is a role for social good of both government and the private sector in research.

Mr. LEWIS. What Dr. Watson is saying is exactly what people said back in the 1970's, when everyone who was in the forecasting business, with one exception, Julian Simon, predicted that the world would run out of oil; and this is why we had to develop energy-efficient technologies at taxpayer expense because we would have run out of oil by the year 2000. The price of oil would be $80 a barrel, $100 dollars a barrel.

Oil is at the cheapest price that we have seen it in years. The price of gas is going down at the pump. This suggests that our energy security is in no jeopardy and that, if anything then, the opposite set of policies would be called for from the ones that were proposed back in the 1970's.

Somehow the same people, though, keep proposing the same policies although with new rationales. Dr. Lashof, for example, referred to a study by the Union of Concerned Scientists to tell us that Kyoto would cost very little or might even actually improve our economy. The Union of Concerned Scientists, however, was one of the leading voices in the 1970's warning us that we were going to run out of oil. Now people are saying that because really oil is so affordable and we are using so much fossil fuel we have a global warming problem.

So what I see here is deja vu all over again, a kind of broken record, in which no matter what the issue is it is always the same set of recommendations: Let's ration energy; let's pour taxpayer money into politically correct research projects.

One of the things that I think is very interesting about the $6.3 billion program is that it would spend a lot of money on work done in a variety of national labs which just happen to be distributed all around the country. I mean this is a form of pork barrel spending.

Chairman TALENT. You are cutting a little close to home now, Doctor. Please confine your accusations regarding motives to the members of the witness panel.

Mr. MICHAELS. We need to correct the record for one slight change. Marlo made a very slight mistake. There was not just one lone voice, Dr. Simon, who argued that we were not going to run out of oil; there were two. The other was a physicist by the name of S. Fred Singer.

Mr. LEWIS. I apologize, Fred.

Mr. LASHOF. I think there are more problems with the record than that, but let me just pick up on one thing.

I was a Member of the President's Council of Advisors on Science and Technology Panel on Energy Research and Development that produced a report last year which was the basis for a substantial portion of the proposed research and development initiative that is included in the Climate Technology Initiative. In fact, that panel proposed a larger initiative on research and development in a variety of energy technology areas than what the administration has actually proposed.

That panel does state, and I fully subscribe to its views, that the challenge of managing the issue of greenhouse gas emissions is the largest challenge that the energy system faces and the most important reason for taking this initiative for it. But there are many others. While oil prices are low now, the fact is that the United States. is more dependent on imported petroleum than we have been at any time in our history, including the 1970's. There are a variety of other pollution problems associated with the way in which we produce and consume energy, all of which suggest that this is a prudent expenditure and well worth it, in my opinion.

Mrs. EMERSON. Dr. Singer.

Mr. SINGER. I go back into history a little bit. I do agree with Marlo Lewis. I remember the days in the 1970's when people were worried, first of all, about running out of oil, which of course we are not doing, and second about our dependence on oil imports.

Now, it turns out that oil imports are increasing and there are countries like Japan that import all of their oil, There are countries in Europe that import all of their oil, and that is not a real problem. But there are people who have been trying to get the Government into energy technology. First they tried to get us into solar power towers in the 1970's, and that did not work. Then during President Nixon, who believed strongly in Project Independence; they made absolutely no sense but it would have put the Government into the energy business.

Then there came somewhat later the problem in Jimmy Carter's administration, when they decided they absolutely had to make artificial oil out of coal. Just imagine what would have happened if the Government had gone into those projects. It was called the synthetic fuel projects. You might remember that, Mr. Bartlett. One of the plants was actually constructed and it has turned out to be completely uneconomic.

I see the same people again trying to push the Government into large-scale activities having to do with energy technology. Of course, the Department of Energy has an ongoing program of research; and we are not arguing against that. But we do not see that the greenhouse situation should be used as an excuse because it does not represent a problem.

Mrs. EMERSON. I thank you and appreciate that. My biggest problem is simply giving advocacy groups grants to do research rather than scientific institutions.

Thank you, Mr. Chairman. I apologize for going way over.

Chairman TALENT. No. I thank the gentlelady for staying through her lunch engagement. I asked her to do it, and I thank her for doing that.

If someone will check to see if this vote is close, I think I am going to keep the hearing going through it. These gentlemen have other things to do. I know Mr. Bartlett has questions he wants to ask, but he is going to go vote and come back. Perhaps I will get in some of my questions in the meantime.

In the first place, let me just, for the record, enter into the record qualifications of some of the witnesses whose motives and their testimony I think has been questioned today.

Dr. Patrick Michaels is a Research Professor of Environmental Sciences at the University of Virginia and is the Virginia State Climatologist. He received his AB and SM degrees from the University of Chicago and a Ph.D. in ecological climatology from the University of Wisconsin, Madison, in 1979.

Dr. Michaels has been President of the American Association of State Climatologists and program chair of the Applied Climatology Committee of the American Meteorological Society. He has published over 200 popular and scientific articles on climate and its impact on society, including a book, *Sound and Fury: The Science and Politics of Global Warming,* published by Cato Books in 1992.

According to the scientific magazine Nature, Dr. Michaels is the United States' most popular speaker on the subject of global warming.

S. Fred Singer, Ph.D., his professional background: Director and President of Science and Environmental Policy project from 1989 to the present; 1994 to the present, a distinguished research professor, Institute for Humane Studies at George Mason University in Fairfax, Virginia; 1989 through 1949, distinguished research professor, Institute for Space Science Technology, Gainesville, Florida; 1987 through 1989, the chief scientist at the U.S. Department of Transportation; 1971 through 1994, Professor of Environmental Sciences at the University of Virginia in Charlottesville, Virginia.

1970 through 1971, you must have just been a stripling during these days, Deputy Assistant Administrator for Policy at the U.S. Environmental Protection Agency. Reads also you have been the Deputy Assistant Secretary, Water Quality and Research, U.S. Department of Interior; first dean of the School of Environmental and Planetary Sciences at the University of Miami; First Director of the National Weather Satellite Center, now part of NOAA, U.S. Department of Commerce; scientific Liaison Officer, U.S. Office of Naval Research; a research physicist for the Upper Atmosphere Rocket program at Johns Hopkins University, the Applied Physics Laboratory. Visiting positions include Research Fellow at the Independent Institute at Oakland, California; Distinguished Visiting

Fellow at the Hoover Institution; guest scholar at the Woodrow Wilson International Center for Scholars; guest scholar of the National Air and Space Museum.

It says here you were a U.S. National Academy of Sciences Exchange Scholar at the Soviet Academy of Scientists.

Dr. Singer, that is a little suspicious. I wonder if they got to you then. That was in 1972.

I could go on. I credit the goodwill and intention of every witness here. I did call the hearing to try to go into the science of this because it seems to me what we are learning in the last few years about what we believed at the time we began moving to what has become the Kyoto Protocol may have been wrong, very significantly.

Dr. Watson, let me ask you. There are a number of ways that we try and predict what global warming is going to be. One of these, I guess, is observation of what the temperature actually is now and has been in the past. That is right? That is one of them, is it not?

Mr. WATSON. It is a combination. The projection for the future is all based on theoretical models.

Chairman TALENT. Just to expedite this because this is preliminary. One of them is observations, and another one is models. Obviously, they are very important in terms of predicting global warming.

Mr. WATSON. Correct.

Chairman TALENT. At the end of the 1980's, the authorities and the IPCC was coming into being at the end of the 1980's, wasn't it?

Mr. WATSON. Yes, in 1988.

Chairman TALENT. As a result of the First Framework Agreement? Was that it?

Mr. WATSON. No. The IPCC was formed before the Convention process. We put science in front of the politics.

Chairman TALENT. There were five or six models that the IPCC relied on that were relied on in developing the protocols that led to the process that led to the Kyoto Protocol. That is right, is it not?

Mr. WATSON. Yes. Probably even more. Probably a dozen or more.

Chairman TALENT. One of those was the model that Dr. Hansen developed; is that correct?

Mr. WATSON. Correct.

Chairman TALENT. Now, you saw Dr. Michaels' testimony about what Dr. Hansen predicted and what has actually happened. Do you agree that Dr. Hansen predicted from, 1988 through 1997 about a .45 degree increase in temperature and that what has actually happened in terms of surface models — well, why don't you put that slide up if you don't mind. Can you do that, Dr. Michaels?

Mr. WATSON. I have not got, obviously, Jim's original paper. So I can neither say that I agree or disagree with that statement. However, by the time of last year, he also published a paper that said, let's look very carefully at all the factors that could change climate between 1979—and the present

Chairman TALENT. I am going to let you get into that, believe me. You are going to have a full opportunity to answer. But surely you are aware if not of the actual down to the detail and number, you are aware that his model for predictions from 1988 to 1997 was wrong.

Mr. WATSON. But that model was not the basis of Kyoto. That model and that projection was not the basis of Kyoto.

Chairman TALENT. I thought one of the reasons I just asked you about the models that were relied on was to establish what the basis was.

Mr. WATSON. It was not that .45. By then the scientists, including Jim Hansen, had realized that CFC's were being controlled by the Montreal Protocol, that the sulfate aerosols were also a factor. So the models that were relied on for Kyoto were in the Second Assessment Report, which Pat Michaels talks about in his testimony, which came down from the high projections you see there to something more moderate. So, the Kyoto Protocol takes into account some of the modifications that are already included in Pat Michaels' testimony.

Chairman TALENT. Let me establish one thing here. Dr. Hansen's model was wrong, wasn't it?

Mr. WATSON. Science evolves, and what we recognize as science evolved is that the projections came down.

Chairman TALENT. So Dr. Hansen's model was wrong.

Mr. WATSON. Dr. Hansen's model was based on the science of the mid-1980's.

Chairman TALENT. Dr. Hansen's model failed substantially to predict what the Earth would — see, I went through the same thing with Dr. Yellin.

Mr. WATSON. That did not influence the Kyoto Protocol.

Chairman TALENT. Dr. Hansen's model did not influence the Kyoto Protocol?

Mr. WATSON. Dr. Hansen's latest models are much more realistic and they were the ones that were taken into account in Japan.

Chairman TALENT. Let me ask you a question. The other models that were extant at the time, Dr. Michaels mentioned five or six of them, were their predictions similar to Dr. Hansen's model?

Mr. WATSON. Are we talking about the ones that influenced the Kyoto Protocol or the ones pre-1990 that influence the Framework Convention? We have got to be careful here.

Chairman TALENT. It is your testimony that these models did not influence the Kyoto Protocol?

Mr. WATSON. No, because by then the models took into account some of the very comments that Dr. Michaels was taking into account in his testimony.

Chairman TALENT. So what level of increase in predicted increase in temperature was relied on at Kyoto?

Mr. WATSON. Exactly as Dr. Michaels says in his testimony, the Second IPCC Assessment Report had a central value of 2 degrees Centigrade with a range of 1 to 3½ degrees Centigrade.

Chairman TALENT. So that model predicts an increase of 2 degrees Centigrade over what period of time?

Mr. WATSON. Between 1990 and 2100.

Chairman TALENT. The range is, like, from 1 to 3, or something like that?

Mr. WATSON. 1 degree to 3½ Centigrade, depending upon what we think the climate sensitivity is and the emissions of the gases themselves.

Chairman TALENT. So, at what level of increase would you think that the Earth was warming dangerously? Would you say that we have a problem if models predict a 1 degree increase by the year 2100?

Mr. WATSON. That is a decision for policymakers to make. For some societies, 1 degree over 100 years may be considered important if sea levels were to rise and threaten them. Other places may feel it is quite acceptable. Unfortunately, different sectors and different regions of the world would feel the effects of climate change differently. That is why it is not a scientific decision what dangerous is, it is a decision for policymakers.

Chairman TALENT. If the Earth's temperature increased by 1 degree over the next 100 years, where would it be relative to the climatic optimum of, say, 500 years ago, around the year 1400?

Mr. WATSON. It would clearly be 1 degree higher. The temperatures today are probably comparable to the temperatures of about 1400.

Chairman TALENT. Does the rest of the panel agree?

Mr. MICHAELS. The climatic optimum was 4,000 to 7,000 years ago and it was 1 to 2 degrees warmer than it is today.

Chairman TALENT. Dr. Singer, do you have a graph or a slide of this graph that you could show?

Mr. SINGER. I think I could do that, yes.

Chairman TALENT. One of the things that has interested me is when you look at what the temperatures of the environment have been in the past you see first that they have fluctuated an awful lot and second that there have been several times when they are higher than they are now and higher than what they are projected to be even under current models. It leads me, and I am a policymaker and not a scientist, it leads me to wonder.

Is there substantial disagreement on the panel about or a disagreement with the panel that that accurately reflects the temperature of the Earth?

Mr. SINGER. This reflects the temperature of the North Atlantic and likely the temperature of the Earth, because we have independent evidence that these temperature changes occur simultaneously around the Earth both at the Northern and Southern Hemisphere.

Chairman TALENT. I would rather avoid a cat and mouse here. I did that with Dr. Yellin.

Mr. SINGER. A thousand years ago, the temperature was higher than it is today according to this data. 3,000 years ago, which is the beginning of historic record, temperatures were very much higher than they are today. The other thing to note is that the temperatures have been changing very rapidly, all of course due to natural causes because there is no human influence that one can think of that would cause these temperature changes.

Chairman TALENT. I wanted the Committee to be aware of this, and I wish there were more Members here to see it. It is not like the Earth's temperature has not increased and decreased in the past. My concern is that it is pretty evident looking at the testimony, which I do not think the gentlemen have refuted, Dr. Lashof and Dr. Watson, that the models used when the world first became aware of the alleged danger of global warming are off, and not just off, they are terribly off by a factor of three-quarters. That is assuming that the surface temperature readings are correct, as opposed to the satellite readings.

The observable data does not justify the models. If you look at it in time of the historical record, there is no basis for concluding from that we are in any danger of global warming and what we are left with, it seems to me, is a whole lot of speculation.

Mr. LASHOF. Mr. Chairman, I don't think that is correct at all.

Chairman TALENT. I would like to be enlightened.

Mr. LASHOF. First of all, this record, as Dr. Singer mentioned, is for the North Atlantic. It is not a global record. The best estimates are that globally over the last 10,000 years it has been at most about 1 degree Centigrade warmer than current conditions. So even at the low-end of the range, we would be approaching temperatures that are unprecedented in the period of written human history over the last 10,000 years. So that is the first point about the historical record.

Chairman TALENT. Does everybody agree with that?

Mr. SINGER. No, I don't agree with that at all.

Mr. MICHAELS. Not at all. The IPCC gives 2 degrees.

Mr. SINGER. Also, if we go about 6,000 years, we will see warmer temperatures.

Chairman TALENT. Is that correct, Dr. Watson, the IPCC says the Earth has been 2 degrees warmer than it is today?

Mr. WATSON. I prefer to give you the answer in writing because I do not want to go just on my memory.

Chairman TALENT. Let me just jump in here and say something, and I guess I am entitled to express a little frustration. That is pretty important, is it not? I mean, shouldn't experts in this field know?

Mr. WATSON. The point of what was 6,000 years ago isn't particularly relevant. What we have to ask ourselves is, if there is a climate change over the next 100 years, how will we, society, react to that climate change.

What we have been trying to do in the IPCC is get the best agriculturists to say, what would be the susceptibility, if anything, to simultaneous changes in temperature, precipitation, and carbon dioxide? What will be the effect on the ecosystem? What will be the effect on human health, etc?

So the key issue is what is likely to be the impact of a climate change and carbon dioxide changes over the next 100 years with the current society and our institutional capabilities today.

Chairman TALENT. It seems to me that one thing you cannot do in this debate is to say that the temperature of the Earth in the past as compared to the present is not relevant to a controversy that is supposed to be global warming.

I can see you are getting frustrated with me, Dr. Watson. I am trying to be enlightened here. If one of the dangers of global warming, for example, is the melting of the polar ice caps and in the recorded history we know that the Earth was warmer than it is today and warmer than it is likely to be under projected models and we know the polar ice caps did not melt, that would seem to me to be a relevant datum. Now, would you tell me why it is not?

Mr. WATSON. It is clearly relevant as to whether the ice caps did or did not melt in the past. The basic point that was just made is that if you take the very lowest projections of climate change, the 1 degree, it — the Earth's temperature — may be comparable or slightly less than 6,000 years ago.

We have not said the West Antarctic ice sheet will melt with a 1-degree centigrade change. What we are showing is a range of projected changes in temperature of 1 to 3½ degrees by 2100. If you go even further out in time, these numbers get even larger, much, much larger; and it is those temperatures that might have the threat of a total destabilization of the West Antarctic sheet.

So, yes, history is relevant for things such as the West Antarctic ice sheet. As far as what the history tells us about agricultural systems, I would prefer to look at the agricultural systems today and say are they amenable to responding to the projected changing climate, also with a population that we are trying to feed today that is very, very different from 6,000 years ago.

Chairman TALENT. Dr. Michaels.

Mr. MICHAELS. First of all, I think that the IPCC, in fact, gives 1 to 2, as you will see in their 1990 report, as the mean temperature change 4,000 to 7,000 years ago with respect to today's temperature; and that, too, is exactly the same number that IPCC gives as their most likely value currently for the change in temperature on out to the year 2100, although, as I have testified, now that we know and Bob Watson agrees with me that methane is stabilizing in the atmosphere and that the effect of carbon dioxide on the temperature itself was overestimated, that median value is going to have to go down.

But before we go further, I want to correct what I think was a substantial misimpression given by two witnesses here and it is an impression that is all over the news; and that impression is that the temperature is warmer than it has been since the year 1400 and that that is the result of human beings' influence on the atmosphere. I want to show you that data if I could.

The study by Mann, et al., and has received a lot of publicity, originates from a large tree-ring record for Northern Hemisphere summer temperatures. This one shows, as you can see, you can see here that, in fact, it looks like this is the warmest temperature since 1400. The IPCC loves to make this statement. The Vice President makes this statement. Everybody makes this statement.

Now, this line here through the data, this smooth line, is nothing but a smooth line put by a statistical averaging technique. The actual temperatures are the up-and-down jagged temperatures, which you can see them drop here substantially around 1550. This record ends in 1980. That is 1980. That click is 1970. 1060. 1950. 1940. 1930. 1920.

All the change in this record took place before human beings could have done it. It is a remarkable misimpression to say that this here is a result of human beings. Just go down the X axis and take a look at when the change occurred. There is 1925. There is 1920.

Chairman TALENT. This is the kind of facts, unless somebody wants to say that it did not happen, that is very disconcerting to me because it tends to suggest, in fact, no relationship.

Now, I understand the precautionary principle and we should err on the side of protecting for these things in case of scientific uncertainty. But we have to have some basis for believing there is a problem before we are going to risk raising energy prices 30 to 50 percent. So tell me why what Dr. Michaels just said is wrong.

Mr. LASHOF. Those are data from the paper, but it is a question of how you interpret the data.

Chairman TALENT. Are you saying the data was wrong?

Mr. LASHOF. No, I am not. I am saying that his interpretation of the data is inappropriate.

The issue is, when you have a system which has natural variability in it and include that there is natural variability in the climate system and then you impose upon that a trend due to some external forcing factor, greenhouse gases, you will have periods where the temperature goes up or down in that record. The issue is, when have you left the range of that natural variability. Because, without greenhouse gases, it is quite likely that the temperature after about 1940 would start going down again. It may be that the increase up to 1940 is due to natural variability. But then you have imposed on that a forcing factor which is preventing it from going back down.

So if you take the record and piece it apart and do not look at the whole thing, you could reach any conclusion you want. That is why we need to rely on a more careful scientific assessment than you have been given here by Dr. Michaels.

Mr. MICHAELS. The last I heard, this is the whole record, Dan.

Chairman TALENT. It seems to me the best that you are saying is that you can interpret this data that does not disprove what you are saying. It still does not prove what you are saying.

Mr. LASHOF. The authors of the paper interpreted this record as demonstrating that, because this record is from proxies and then you put on the end of that actual measurements, that with a 99 percent statistical probability, 1997 is the warmest record since 1400. That is what the authors of this paper that was published said.

Michaels wants to make a different interpretation. That is his right. That can be debated in the scientific community. I want to make that point.

Chairman TALENT. It is not a question of interpretation.

If it is the warmest in record, it is still true in that, I think irrefutably, that the warming that got us here occurred in the early part of the century.

Mr. LASHOF. But what is not shown is the warming after this proxy record ends. Well, you can see it in the dash line up to 1990; there is continued warming there. There is greenhouse forcing before 1950, as well, as Dr. Watson pointed out before. It is modest in that period but it is not absent.

So when you look with models that reconstruct the past and compare that with that record, there is no inconsistency. So the suggestion that this shows that the models are wrong because the warming took place during the wrong period just does not stand up to a rigorous analysis. It looks good when you throw these charts up and wave your hands, but the rigorous analysis that has been done suggests something different.

Chairman TALENT. My understanding of the models is that when applied in a predictive fashion, say for the last 10 years they were wrong, when applied over the last 100 years they were wrong. Now, is there any context in which they have been right?

Mr. LASHOF. The statement you just made is incorrect. The model that Hansen published in 1988 predicted the warming over a 10-year period. That is certainly true.

Chairman TALENT. By a factor of 3; is that correct?

Mr. LASHOF. Yes.

Chairman TALENT. So you are telling me at least that that model was wrong?

Mr. LASHOF. First of all, the paper was not intended to make a short-term prediction. I think that is important. It was not intended to look at only a 10-year period. Because over a 10-year period there is a lot going on, including volcanoes and other things. So that prediction was wrong, yes, I agree with that.

Mr. LASHOF. But the models that are currently being used that were used in the IPCC Second Assessment Report, if applied over the last 100 years, are correct. They have excellent agreement with the data over that period.

Chairman TALENT. I am sorry. Repeat that.

Mr. LASHOF. The current models that are used by the scientific community that were used in the 1995 report of the IPCC, if applied over the last 100 years, have excellent agreement with the surface temperature.

Chairman TALENT. Now, that would seem to me to be significant. Is that correct?

Mr. MICHAELS. The problem is the way they were corrected. The correction that is in vogue is to say that sulfate aerosol, which is another emission that goes in the atmosphere along with carbon dioxide, is the reason that the warming did not appear.

Again, on page 4 of my testimony, I have the record since 1979 for the Southern Hemisphere temperature, free atmosphere temperature, which is a very sulphate-free atmosphere. It is crystal clear, and there is a significant drop in the temperature that sulfates could not have had anything to do with.

Further, and very importantly, the landmark study upon which the sulfate hypothesis is based used data from 1963, which is when Mount Agung went off and cooled the atmosphere, to 1987, which was the warm, last warm point in the record prior to 1997. When you take a look at all the data, which begins in 1958 and ends in 1995 and 1996, depending upon what set you have, you do not see any change at all.

Chairman TALENT. Let me ask this question, Dr. Michaels. If this model as currently corrected were applied to the warming throughout the 20th century, would what it had predicted for that correspond with what we know actually happened?

Mr. MICHAELS. Which model?

Chairman TALENT. The model as currently corrected by this aerosol correction.

Mr. MICHAELS. You could make it do it, but that is not the point. You see, you can make a model drop its temperature from 1940 to 1970, which is what the Earth did, by just assuming somehow there were more sulfates in the atmosphere. All you will do is write an equation that reduces the amount of radiation and that radiation that hits the surface. That does not say that is what caused it. In fact, it makes you right for the wrong reason.

By the way, James Hansen says that over the course of the last two decades the sulfate influence on the climate is zero, that there is no sulfate cooling being exerted in the atmosphere. What is going on here?

Mr. LASHOF. Within that period. That is the whole problem with your analysis is you take a period where there already have been substantial forcings from both greenhouse gases and aerosols and just try to make a comparison from 1979 to the present.

Chairman TALENT. I want Dr. Watson to comment. Then I am going to recognize Mr. Bartlett.

Mr. WATSON. Jim Hansen published a paper on November 27, 1997, to try to explain the 20-year period, 1979 to 1996; and he said, if you took into account all the factors that we know about, such as natural variability of volcanic eruptions, the El Nino phenomena, plus take into account ozone depletion, as well as taking into account changes in greenhouse gases, he gets basically a model projection of temperature change about .2 degrees centigrade, which is very close to observations. He does not even have to invoke the sulfate effect over that short period.

This model shows how there was a consistency between theory and observation Sulfur was not certified during the period 1979 to 1996, because many countries in the Northern Hemisphere, Europe, etc., started to decrease Sulfur emissions because of the acid deposition problem. So it is over the longer time period that you have to take into account the effect of the sulfate aerosol.

So quite clearly, what you have in front of you is a number of scientists that published, who do believe their explanation is reasonable, and obviously Pat, who has a different opinion.

Chairman TALENT. I am going do recognize Mr. Bartlett now. Thank you for being patient.

Mr. BARTLETT. Thank you very much.

Let me apologize for my absence during the last testimony and during much of the questioning. We were trying to move the ball a little further in bringing some sanity into basic recruit training where they now are sleeping next to each other and pretending that there are not physical differences between men and women.

Listening to the discussion here, I was reminded of an exchange. Just about 50 years ago now, I had just completed my master's in science and was working on my doctorate and my major professor 1 day said to me, "You know, Roscoe, there are 3 kinds of liars." That is 3 degrees of liars. "There are liars, and there are damn liars, and there are statisticians." He was kind of exasperated with what statisticians can do with data.

It is very clear from these discussions that he who frames the question determines the answer, that you can make the scientific data march in any direction that you want them to march. There is, and I wonder if you will agree or disagree with me, a general misconception of what science does. Science does not prove things or disprove things. Science tests hypotheses and they find a statistical level of probability that the hypothesis is correct or incorrect, that we do not really prove or disprove things.

Is that a correct statement of what science does?

Everyone is nodding in agreement, let the record show.

I would gather that, if you were to summarize, the present status is that there is general agreement on what the data are, there is great disagreement on what the data indicate.

Let me comment for just a moment on the renewables and energy efficiency, because I happen to be a strong proponent of renewables and energy efficiency for very different reasons than trying to cope with global warming. We have 2 percent of the world's known reserves of oil. We use 25 percent of the world's energy. From a national security viewpoint, one just has to look at that I think.

It is clear that fossil fuels will not last forever. We are not making them at the rate we are using them for sure. When they run out is problematical, and I have a major concern that we have a very large petrochemical industry that really makes the point that gas and oil are probably too good to burn. We live in a plastic world; and, essentially, all of that comes from oil.

So I am a very strong proponent of renewables and energy efficiency, primarily because I think that, as a matter of national security, that we cannot be increasingly dependent on foreign oil. As long as we are the umbrella under which nations like Denmark and Japan exist, they are OK. We do not live under anybody's umbrella; and, so, I have a concern for that.

Let me ask you a question. Carl Sagan was concerned that if we had a nuclear exchange that we would have a nuclear winter. That is because all of the pollution thrown up. There is some justification for this. I do not know the extent to which it is justified. But when a large volcano goes off, we have some temporary global cooling.

Now, you cannot produce more greenhouse gases without producing other pollutants. My question is how do we know what that balance will be? How do we know that when greenhouse gases increase and other pollutants also increase that the net effect is going to be a global warming or a global cooling?

Mr. LASHOF. Mr. Bartlett, let me try to address that. You are right to draw that analogy because the cooling effect of nuclear weapons is similar to the cooling effect of aerosols. The key point is temporary. The sulfate emissions which are associated primarily with burning coal, also high sulfur oil, there is almost no sulfur in natural gas. So if you burn natural gas, you produce CO2, you do not add any cooling sulfates to the atmosphere.

With coal and oil it depends on how you burn them and what technology is used. If you allow the sulfur to be emitted, you have very serious both local and regional pollution problems, direct inhalation of particles as well as acid deposition. So, for that reason, most countries around the world are controlling sulfur emissions while currently allowing fossil fuel burning to continue. The sulfur is on the order of 1 to 2 percent of the content of the fuels. It is not an inherent part of the combustion process. Whereas, carbon dioxide emissions, that is how you get the energy out of fossil fuels. There is not really any way to burn them without producing carbon dioxide.

There is some interesting ideas that are in the research area about ways to maybe take that carbon dioxide and reinject it into geologic deposits, and that is certainly worth looking at. But right now we do not have any technical means to do that. But we do have the technical means to take the sulfur out, and we are doing that more and more.

The other key point in thinking about this is that the sulfur emissions, once they are emitted into the atmosphere, they stay in the atmosphere for about 14 days, 10 to 20 days roughly; whereas, the carbon dioxide has a residence time in the atmosphere of, on average, about 120 years.

So while you may have, and I think that is what we have seen over the past, for a period of time, the sulfate cooling offsetting the accumulation of greenhouse gases over some period of time, eventually that sulfur will fall out of the atmosphere. We will stop emitting it and then it falls out of the atmosphere very quickly; and what you are left with then is the greenhouse gases, which you are stuck with for a very long period of time. This very long timeframe is why I think Dr. Lewis's argument that you can apply the precautionary principle equally to the economic sphere as you can to the environmental sphere is wrong.

In one case you are dealing with what is essentially an irreversibility of putting greenhouse gases into the atmosphere, you are stuck with them for generations. If you start moving down the path of reducing our reliance on fossil fuels, you have an ability to adjust your policies over time as you learn from how those activities take place.

Mr. LEWIS. I think the fact that carbon dioxide has a long residence time is an argument in favor of a wait-and-see approach rather than a let's-take-action-now approach. In fact, one of the leading modelers in the field, Tom Wigley, along with Richels and Edmonds, in a 1995 article in the journal Science, argued that we could wait until the year 2020, I believe it is, before making the kind of reductions in greenhouse gases that Kyoto calls for and the only difference it would make by the year 2100 is two-tenths of a degree centigrade.

So the long residence time of CO2 means that if we start making reductions in greenhouse gases now, it is really not going to make a big difference in the year 2100 compared to if we wait a decade or two. So I do not think that this is an argument for precaution. I think it is an argument for prudence, which is, we do not have to be in a hurry, so let's not rush.

Congressman, I wish I could make a few comments on your enthusiasm for renewable fuels. I share your concern about America's security. Obviously, we are the leader of the free world; and, so, we have to take things into account other than mere economic considerations.

I would suggest that one of the best things we could do to diversify our sources of energy, fossil fuel energy, should we need them, would be to lift the restrictions on drilling on the Continental Shelf. I mean, everybody I think who is in this field who is honest will acknowledge that there is more danger of an oil spill from a tanker bringing imported oil in than there is from a platform offshore.

So if the environmental community is so concerned that the world is going to run out of oil or that America is too dependent on Middle Eastern oil, why do they oppose drilling on the Outer Continental Shelves? Why do they oppose any kind of drilling in Anwar, when the Audubon Society, for example, on its own wildlife range in Louisiana allows oil companies to drill there. I mean, it can be done in an environmentally sensitive way.

I would also point out that the percentage of oil imports as a share of the overall merchandise trade of the United States has gone down dramatically since the 1970's. So, yes, we are importing more oil, but that is because our economy is bigger, we are doing more of all kinds of things. But as a percentage of the economy, oil imports are declining.

Finally, I would just note, actually there are two more points I would like to make if I could, is that the notion of dependence on foreign oil I think is not correct economics. The reason why Americans import oil, the reason why we buy it abroad is because we value the oil more than the dollars we are sending out. I mean, this is just another kind of trade. We also buy cars from abroad. We also buy rice, even though we are one of the world's biggest exporters of rice. The gains from trade are mutual. That is why it occurs. So the fact that we are buying some oil from some foreigner I do not think is in itself anything to be worried about. It is also interesting that a lot of that oil is not coming from the Middle East, it is coming from the North Sea, it is coming from South America.

Just one last point, which is that, for two decades, taxpayers and rate payers have subsidized solar power and wind power to the tune of billions of dollars, and these technologies are still two or three times as expensive as the most competitive energy from fossil fuel. I think really enough is enough. If they cannot make it on their own in the marketplace, we should not ask more taxpayers and consumers to pick up the bill.

Mr. MICHAELS. If I could, your question which Dan alluded to talks about a balance between putting warming things in the atmosphere such as CO2 and sulfate. When it became apparent that the model of the type that I began my testimony with had failed, the United Nations recognized this and they wrote in 1995 the following: The IPCC said that "greenhouse gas models, using only changes in greenhouse gases, tend to predict more warming than has been observed unless a lower sensitivity," that means it will not warm as much, "is assumed, given an increment of carbon dioxide." Then they say, "there is growing evidence that increases in sulfate aerosols are partially counteracting the warming."

That is an expression that either there is something wrong with the sensitivity estimate of the climate itself or sulfate aerosols are countering the warming.

Now, the Chairman asked me, if you took a model like that, could you simulate the past. I said, sure, you could tweak a model to do it.

Would you like to see what should be the easier answer on this sulfate greenhouse problem? Can you simulate the present? Do you think we can? Do you think we can take a climate model and simulate the current climate? Would you like to see?

Mr. BARTLETT. Which model is this?

Mr. MICHAELS. National Center for Atmospheric Research. The most sophisticated model we have, as a matter of fact. It has just hit the literature.

The slide I want to show you, this slides shows the difference between the known temperature in July, the average temperature at every place on the planet in July, and what the model says the July temperature is.

Now, let's take a look here when Congressman Talent is from, St. Louis. That is an area of 11 degrees Celsius. It says that St. Louis and the area around it is 20 degrees Fahrenheit warmer than it really is. Now, why do we not see errors like this when people talk about how much the climate is going to change? Because when we do the process, we take a model like this with these errors, double the carbon dioxide, or gradually increase the carbon dioxide, and look at the change between one model and the other having subtracted these massive errors out.

But, in fact, these massive errors mean you really cannot simulate the present climate. So I think that any modeled comment when sulfate aerosols are going to be overcome by greenhouse gases or even if sulfate aerosols are the reason that it did not show up is at best premature, especially given the fact that the entire southern half of the planet shows a cooling in the satellite record over the last 2 decades when there is no sulfates to cool it.

So that is my answer as to when one will cross over and cancel the other.

Mr. SINGER. Let me back this up in a simple way. The aerosols have been brought in by the IPCC as a Band-aid or as a way of papering over the fact that the models cannot explain the past climate record. They will disappear. I guarantee you that in the next IPCC report you will not see the aerosols in the way you saw them in the last one.

Mr. BARTLETT. Let me make a brief response to the question about Anwar and offshore drilling. I have been to Anwar. I do not have the least question that there is any environmental insult that would occur if we were to drill there. The footprint is very small. They do their activity in the wintertime, where they make the roads out of ice. When the summer comes, there is no evidence a road was ever there. They are very good at not disturbing the environment.

That does not mean that I am supportive of exploiting the reserves offshore and in Anwar. The reason for that is that I think we need to husband what oil and gas remains in the world. It is the basis of a very large industry, the petrochemical industry.

I am reminded of a story of a now extinct ancient kingdom where their sustenance came from a lake and the king's advisors came in and they told him, "King, there is a very interesting thing happening on the lake. There is something growing on the surface of the lake and it is shading the lake." They did not know the food chain, but they did know that there were not any fish there. That is what they lived on. So the king asked, "How big is it?" And they had impressed on the king that when it covered the whole lake, they would have no sustenance and they would disappear. So the king says, "How big is it now?" They said, "Well, it is not very big. But it is doubling every day." The king said, "How much of the lake do we need for our sustenance?" And they said, "Well, we could get by with a half." He said, "Let me know then when the lake is half covered." Of course, the next day, if it is doubling every day, it would be totally covered. So, that is my concern about gas and oil.

I do not know if you are familiar. He is no relative of mine. Albert Bartlett from the University of Colorado gives the most interesting 1-hour lecture that I have ever heard on the biggest failure of our industrialized society, and that is our inability to appreciate the exponential function, and he makes a pretty compelling argument that we have been on the down side of pumping oil in the United States since 1970.

As a matter of fact, I think that essentially every year since 1970 we found less oil and pumped less oil than the year before; and that is because there is less oil there to be found and pumped than the year before.

Even if we do have global warming, my next question is, what difference does it make? Through the past we have had very big temperature changes on our climate. We have had glaciers that have come down into southeastern Ohio. We are still finding mastodons frozen in the tundra, frozen as a result of a precipitous temperature change, because they still have subtropical vegetation in their stomachs. So we have had very big temperature changes in the world.

Let me ask simply about the time when those mastodons lived where there was subtropical vegetation growing in what is now perpetual permafrost. What happened to the polar ice caps then?

Mr. CHRISTY. I do not know that the vegetation was subtropical in the sense of what I think was subtropical in those northern regions.

Mr. BARTLETT. I am simply repeating what I have read in the literature that occasionally we still find these mastodons and they have what we would describe as northern Florida type of vegetation in their stomachs.

Mr. CHRISTY. That is a little unusual, because a mastodon to me is a very large warm-blooded animal that is built for cold climates. So, the climate change would be what would cause that, I can understand that was rapid. I had mentioned an example of a climate warming from the Pleistocene period that was extremely rapid, 11,600 years ago. So a warming like that could have an effect such that it could kill off a population of mastodon. I suspect that human involvement might be there in the sense that our ancestors hunted that critter.

Mr. BARTLETT. There was several years ago in *Science* a whole series of articles that took up much of one issue talking about the mastodon phenomenon; and the conclusion was that we could not from any of our models explain how that could have happened.

Let me suggest a scenario. If we have increased greenhouse gases, the temperature increases. This temperature increase increases evaporation. This increased evaporation produces more rainfall. Now, you have two things there that are increased that much benefit the production of food. One of them is an increase in CO2. Plants are pretty efficient. They make do with .04 percent CO2. We have to have about 10 percent oxygen. So we are not very efficient as compared to plants. They do much better, by the way, with more than .04 percent CO2.

Now, why a world in which we have more plant growth and more water to support that plant growth, why isn't that maybe a better world than we have today? Truly, if we have some warming in the world, it would be different. But you would have a tough time convincing southern South America and Siberia and northern Canada that a little global warming would not be a good thing.

As long as you have these two things occurring together, more rainfall and more CO2, therefore, a priori, how can you say that even if there is global climate change and warming that the world would be different and how could you say that it would be worse?

Mr. WATSON. As I said in my testimony, if one looks at what IPCC projected for agriculture, it actually produces an increase in agricultural production in North America and also in Siberia, which offsets a decrease in projected agricultural yield in the tropics and subtropics.

The tropics and subtropics, at least in most of these models, are projected over the land areas to become drier insofar as even those areas with the same precipitation, the increased temperatures lead to more evapotranspiration. So literally all of the models that we have used for agricultural production take into account both climate change and the fertilization effect of CO2 and those models show an increase in productivity in the high latitudes, a decrease in productivity in the tropics and subtropics, and in the end, almost no effect on the global productivity.

But there is a change in the regional distribution of agriculture. If you believe that the food trade system will work perfectly, you could become complacent. But the trouble is that these projections are showing a significant decrease in productivity in those countries, unfortunately, where there is hunger today.

Mr. BARTLETT. But isn't it inescapable, unless the increased rain and the warmer temperatures included a lot of nonarable land, that the inescapable conclusion is that with more CO2 and more rainfall you got to have more food production? I do not see how you can escape that.

Mr. WATSON. But parts of the world do not have more precipitation. Globally averaged, there will be more precipitation.

Mr. BARTLETT. Of course. If you have more evaporation, there will be more precipitation. It will not stay up there forever.

Mr. WATSON. Right. But it does not mean all parts of the world will equally become wetter. Some will become wetter. Some may become drier.

Also, these crop models do not bring into account the potential increase in pests due to higher temperatures. This is an area of research where one has to look at it very, very carefully.

Mr. MICHAELS. Mr. Chairman, I think we should quote the IPCC, because I think Bob is sort of dancing around about what their statement is about this, and I just want to show you what they say.

Here is what it says. Read that for yourselves.

Mr. BARTLETT. Yes, what this says is that the world will change. It will be different. There will be less rainfall some places, more rainfall other places.

If the increased rainfall was in varable lands with more nutrients than the land that we have been wearing out for the past couple of thousand years, wouldn't that be argued to be perhaps a good thing for the world?

Mr. MICHAELS. I would agree it is very clear that the scenarios of agricultural disruption, Bob is right, they do generally predict more moisture to occur. But you see, they argue that there will be more evaporation also and, therefore, there will be more droughts and/or floods, whatever that means.

Now, I think we need to ask the question when we see statements like that what is really happening. There are two scientists at the U.S. Geological Survey, Harry Lins and a Mr. Slack, who recently gave a paper in which they showed that, our rain gauge network is not really very good. Rain is pretty spotty stuff. But stream flow is a wonderful indicator of rain — and they found that the highest category of flow, which are the floods, show no change, contrary to what we have heard, and there has been a decrease in the frequency of the lowest category, i.e. droughts, a decrease.

So what you are saying turns out to be true, we are creating a world that is this fact a better moisture climate for agriculture than the one that we have now.

Mr. BARTLETT. Mr. Chairman, it is obvious that from both a layman viewpoint and from a science viewpoint we could discuss this for a very long time. But I think it is very clear, and I am really appreciative of the witnesses who have come and their testimony. I think it is very clear that we are a long way from any consensus on the data that would justify any massive disruption of our society, and that is what they are proposing. It would be pretty massive.

We are here on the Small Business Committee today, and I am also on the Science Committee, we are here on Small Business and the reason that we are here is that because the changes that are suggested as a result of the Kyoto Protocol would in many cases be disastrous to small business.

If the future of our society was dependent on these changes, then one would say, that you just have to tighten your belt and do it because mankind depends on it. That just is not the case. We just are nowhere near from a scientific viewpoint, and I come from a science background, of any consensus on this to justify any action that would be disruptive in our society. These things occur very slowly. We have plenty of time to wait and watch and take prudent action.

I thank the witnesses, and thank you, Mr. Chairman.

Chairman TALENT. I thank the gentleman. The discussion has gone for a very long time. Let me go ahead and ask a few more questions. Then I hope we can wrap it up.

True or false. The models originally used to predict global warming were wrong, at least in that they did not include the effect of aerosols? Is that true or false in your judgment?

Mr. LASHOF. True.

Chairman TALENT. True or false. The Earth's temperature is not warming at a rate consistent with the initial predictions even according to surface temperature measurements?

Mr. MICHAELS. True.

Chairman TALENT. The Earth's temperature is not warming at a rate consistent with the initial predictions, even according to surface temperature readings?

Mr. WATSON. With the initial predictions but they are not the ones used for the Kyoto Protocol.

Chairman TALENT. I understand that. So that statement is true?

Mr. LASHOF. I am not actually sure whether that is true, because I am not sure the record from the time that those predictions were made to now is long enough to make it in a rigorous way. In other words, there is enough natural variability over a one-decade period that I am not sure you could say statistically that the temperature record is inconsistent. So I am just not sure.

Chairman TALENT. If there is enough variability over a one-decade period, then what is the relevance of statements like "each month so far this year has set a new all-time record for global near-surface temperatures. June 1998 exceeded the previous record warmer June 1994 by nearly .4 degrees"?

Mr. LASHOF. Because 1997 and 1998 are warm years as part of a series of very warm years over the last two decades.

Chairman TALENT. There is no relevance in that statement, is there, Dr. Lashof?

Mr. LASHOF. I believe it is a relevant statement.

Chairman TALENT. I mean statements about what happened this year or statements about a flood someplace this year or a fire someplace this year are irrelevant to this debate, aren't they?

Mr. LASHOF. Statements about individual fires or floods I included a couple of references in my testimony to illustrate the fact that contrary to some claims, we as a society are still very vulnerable to climatic events, not to demonstrate, and I think I was clear about this in my testimony, not to demonstrate that as a result of, say, the fire in Florida that that proves that global warming is taking place. I did not include that in my testimony for that reason.

Chairman TALENT. You put it in your testimony to show that we are vulnerable to fires or we are vulnerable to heat waves or we are vulnerable to floods? I mean is that a subject of debate?

Mr. LASHOF. I believe it in fact has been debated here. Pat Michaels said the climate could warm up a lot and it does not matter much. I think that is not true.

Chairman TALENT. Wait a minute. Nobody is arguing that we are vulnerable in the event of heat waves or vulnerable in the event of floods. The question is, A, whether the Earth is warming. I am just going through a series of true or false statements to show that there is nothing in the record that I have seen so far to indicate that it is, although more than slightly. Second, if it is warming, is that warming connected to these climatic events?

The point I am making is, the existence of these catastrophes or these heat waves or the temperature in a given year or a given month is not relevant to those issues. It is not relevant to whether the Earth is warming and it is not relevant as to whether warming that is occurring is causing those events? Now, is that true or false? I will ask everybody.

Mr. MICHAELS. True.

Mr. LEWIS. True.

Chairman TALENT. Dr. Watson, is it true or false?

Mr. WATSON. I think the point one is trying to make is that if indeed you get a long sequence of unusually warm months, you have had to ask the question is that statistically expected or is it the type of event that might become more frequent in a warmer world. So, any single event cannot be attributed to global warming. But if one starts to see a sequence where you see in the probability of these hot events there are more and more, it is the sort of thing to note to society that we appear to be changing our climate.

Chairman TALENT. I would like to read you a statement. "It is important to remember that short-term decadal natural fluctuations are expected. Hence, the key issue is the trend in surface temperatures over the longer term that averages out these short-term natural fluctuations."

That is from your statement for the record, isn't it?

Mr. WATSON. I agree, and that is not inconsistent with what I just said.

Chairman TALENT. It just seems to me that you just said that if there is a few years in a row that they could be indicative of a trend. Well, they could not be indicative of a trend too, right?

Mr. LASHOF. But a trend is observed and the most recent data is consistent with that trend continuing. So I think that is relevant information.

Chairman TALENT. Trends we are observing are at very best much less than what was originally predicted. We know we have the data for what is happening in the last few years. The surface temperature readings are at an increase of what, Dr. Christy?

Mr. CHRISTY. .15 per decade, so about a quarter degree.

Chairman TALENT. The satellite temperature readings are actually showing a slight cooling. So the trends insofar as we can observe them are against the hypothesis that supports global warming.

Mr. WATSON. No.

Mr. LASHOF. Not true. The trends are consistent.

Mr. CHRISTY. Let me say that is not correct. The trend in the troposphere is not consistent with these issues. Every climate model that we examined had a greater warming in the troposphere than it did at the surface, every single one. We examined 14. So, this phenomenon is just one of those unexplained ones that, hopefully, further climate research will nail down and give us a better idea of what is happening in the atmosphere.

I would like to say, I hope in my testimony I made the point that in this country, in the data sets I showed, that these extreme events are not increasing.

Mr. MICHAELS. I want to underscore that. We looked in Climate Research in the April 9 issue for increasing trends in record high temperatures. We could not find one. That was in the IPCC's data. To try and say that a hot month means global warming or a hot day, we clearly do not see it.

Chairman TALENT. I am going to take a stab at trying to get some certainty here, and then I am going ask a question or two to explain why I have some passion about this and then I am going to close it off.

Dr. Christy testified that if one divides the recorded climate history of the entire U.S. into halves, and by the way, the recorded climate history since when, 1880 about?

Mr. CHRISTY. About 1895.

Chairman TALENT. All right. In 1940, and I guess that would do it, about half the recorded history was before 1940 and about half was after, and looks at all-time high temperatures, one would see that four States had equaled high in both halves; of the remaining 46, 35 recorded their all-time record high temperature prior to 1940 and only 11 have since 1940.

Now, is that evidence relevant to show that the Earth is now warming? What does that evidence show, Dr. Christy.

Mr. CHRISTY. To me that evidence just shows that you can spin statistics any way you want. I do not think it shows either way what is happening.

Chairman TALENT. The thing that I have seen of all this that I am the most suspicious of are statements or comments that rely heavily on there was a drought here or a flood here or it was hot here or it was not hot here, because I do not think that a responsible scientist would say that that is evidence of anything. I take it you all agree with me on that?

Mr. LASHOF. It is evidence of something but it is not evidence in and of itself.

Chairman TALENT. Not evidence of global warming. So a drought someplace in one particular year is not evidence of anything, right?

Now, all evidence that taken in the totality, the number of these extreme events is not increasing. That was your testimony, Dr. Michaels, for the record.

Now let me explain why this is so important to me. We are in the middle of this muddle, and it looks to me from looking at this evidence that it does not support a global warming hypothesis or the warming of the Earth to a degree that anybody can say with certainty is of any harm to people. We are talking about this manifesting itself 100 or 150 years down the road. In order to prevent it, we are talking about the implementation of a protocol, which I believe after our hearing with Dr. Yellin could raise the price of energy in this country anywhere from 30 to 40 to 50 to 60 percent.

So to me there is a certain irony in debating about what is going to happen to the production of energy someplace when there are real world people out there, and I know what is going to happen to the production of agriculture and food for people if the energy costs in this country go up 30 to 50 percent.

How can we talk in these kinds of terms of hundreds of years down the road when there are people who are not going to be able to heat their homes, when the farmers are going to go out of business, when the people are not going to be fed as they should be. If the United States cannot produce food because energy prices go up, what happens to the rest of the world now, not in the year 2100.

This is why I feel such passion about this. You are predicting something which may or may not happen and on the basis of evidence that to me falls very short of proving any really serious harm. But we know what happens in the here and now to these people if we do this to the price of energy. Why aren't you more concerned about that? Go ahead and tell me why I am crazy.

Mr. WATSON. I would say there are four factors that I would take into account if I were a policymaker, which I am not. First, the changes are not necessarily going to be slow. They could be nonlinear. We were very, very surprised in the stratospheric ozone issue where we thought the changes would be very slow over the next 100 years and we found there was a nonlinearity in the system, especially in Antarctica, but now we have also found it in the Arctic.

So, in other words, slow changes may occur but there also could be some thresholds and nonlinearities, which we do not know. We have to have that in mind.

Second, the lifetime of atmospheric carbon dioxide is so long, that effectively, if we wait for cause and effect to be established and we do not like that changed world, we cannot reverse it for decades to centuries to millennia as my figure 1 shows.

Third, it is not clear to me that these mitigation costs are going to be that exorbitant. The IPCC argues that indeed there are many approaches through using energy efficiency and using fuel switching where we can start to make progress to a different energy future that would not be as expensive as some people say.

Indeed, Shell and British Petroleum have talked about some plausible futures where there would be a significant shift over 5 decades. We just have to start now to having a different position effectively in 50 years time.

Also with articles 6, 12, and 17, which are the flexibility mechanisms in the Kyoto Protocol, there is project-based emissions trading (6 and 12), and emission rights trading (17) like the sulfur market in America, one effectively does not have to take all those hits in the United States where the cost of abatement may be higher than the cost in China. One could effectively allow the market to find the cheapest options to reduce greenhouse gases. Therefore, I am very skeptical about these very, very high costs that some people claim, and so is the IPCC.

Chairman TALENT. Dr. Lashof, do you have a comment?

Mr. LASHOF. I agree I think with everything that Dr. Watson just said. I do not know why if we have spent the morning talking about uncertainties in atmospheric models that are based on physics and real data and careful assessment and said, yes, there is some uncertainties about that, although I believe they demonstrate a very serious risk that we are facing here, that we would take at face value economic models which have been clearly, developed and run with assumptions directly paid for by fossil fuel interests to show exorbitant costs, and I just do not believe those numbers.

I think there is clear evidence that we can implement the Kyoto Protocol, which is indeed a very modest first step in starting to curtail greenhouse gas emissions, without harming the economy and that it is a prudent thing to do, it is the right thing to do. I think it is in the national interest.

Thank you.

Chairman TALENT. Would either of you care to offer an opinion on whether you would do it if indeed it does raise energy prices by say 30 percent? Do you think the scientific justification is there to support that kind of cost increase?

Mr. LASHOF. I think you have to look at the fast that energy prices are only one factor in a very complex economy. If it were to cause a 30 percent reduction in our gross national product, I would certainly not support moving forward. But if we, for example, as Congressman Bartlett suggests, want to husband our use of fossil fuel resources and decide that we should perhaps pay a little bit more for those fuels in order to encourage efficiency and use those resources to reduce other taxes, for example, other taxes that farmers pay or other taxes that people pay, I believe that that may be a prudent thing to do. So I think energy prices as a sole focus is inappropriate.

Chairman TALENT. You all do not have the benefit of the hearing which we held when Dr. Yellen was present, and there was anything but clear evidence that this protocol would not have that kind of an impact. As a matter of fact, she was unwilling to release the basics of her economic analysis or even to answer the simple question about what the impact of reducing greenhouse gases below 1990 levels would have on the price of a gallon of gasoline apart from this trading scene. I came away from that with much less confidence than you all have.

I will recognize the gentleman for a final question.

Mr. BARTLETT. Thank you very much.

I would just like to note that the extremes of high temperatures or low temperatures simply tend to indicate the shape of the bell curve. All these temperatures fall under a bell curve. If we have big extremes we have a flat bell curve. If they are not big extremes we have a more pinched bell curve. But the extremes do not indicate anything about the center of the bell curve, I think is the point that you all were making.

I have a question about CO2 in the oceans. CO2 is very soluble in water as compared to oxygen, by the way, something like 40 times more soluble in water than in oxygen. Since the whole food chain in the ocean begins with phytoplankton and since phytoplankton rely on CO2 for their growth, two questions: First, wouldn't some increase in CO2 be good for food production in the oceans because it would feed the phytoplankton?

Mr. SINGER. Clearly, yes. This, in fact, opens up a very exciting possibility that has not been widely discussed: Instead of cutting emissions of CO2, if one wants to reduce the concentration, one should instead sequester the CO2, take it out of the atmosphere, and put it into the ocean.

The way to do that was proposed by an oceanographer John Martin a while back, and this proposal has now been tested experimentally, that the ocean be fertilized with a micronutrient, that if iron is released in the ocean, this would drastically increase the growth of phytoplankton, which of course would draw CO2 from the atmosphere.

This has now been verified, tested, published 2 years ago in Nature, and is now the focus of ongoing work and I think is a very promising avenue of research, not because we are concerned about greenhouse warming but because we are concerned about feeding people, who need more fish.

Mr. MICHAELS. Along the lines of your question, I would like to quote again Hansen, who really kind of started this thing off 10 years ago. In his most recent paper when he was commenting upon the fact that CO2 was not increasing at the rate that people thought it would, he says, "apparently the rate of uptake by CO2 sinks, either the ocean or more likely the forests and the soils, has increased."

So I think there is a general agreement that we are sequestering more CO2 as we increase the concentration in the atmosphere.

Mr. BARTLETT. My question, then, why isn't that good both on land and sea?

Mr. MICHAELS. It is a good thing.

Mr. BARTLETT. One of our Members disagrees.

Mr. WATSON. The first question you have to ask is, is ocean productivity limited by carbon dioxide. Fred Singer's own answer shows the answer is no, it is probably limited by other nutrients such as iron. So more carbon dioxide into the ocean may or may not increase productivity. It is more likely that productivity is limited by sunlight and by other nutrients such as iron.

In fact what he said about the iron fertilization, while it may be correct, you would have to see what the other effects on the ocean were. The fact that iron fertilization works shows it is not carbon dioxide limited.

As far as the terrestrial biosphere, I think it is indeed correct to say that the terrestrial biosphere is indeed absorbing carbon dioxide. It is almost certainly in the northern latitudes at the moment. The question is whether that continues in the future, and we do not know. If we do not understand what the process is, was it simply regrowth of forests after we cut all the forests down in the Northern Hemisphere, or much of the forests down in the Northern Hemisphere at the beginning of this century versus is it a carbon dioxide effect? We cannot project what is going to happen in the future, will the CO2 keep being absorbed by the terrestrial biosphere, especially as we are cutting the tropical forests down. So the extrapolations made in Pat's paper need to be very carefully examined.

Mr. LASHOF. More nutrients is not necessarily a good thing when added to a natural system. Mr. Bartlett, just one example I think you are well aware of, just look at the Chesapeake Bay. Its biggest problem is not too few nutrients, it is much too much nitrogen in that case, which is causing huge ecological problems, complete collapse of various fisheries, very serious problems.

So when you are talking about agriculture, that is one issue, that you want to put more nutrients in it to grow it faster. No question about that. Dr. Watson talked about other problems. You may not have the soils and other things. But if you are talking about the natural world, the assumption of adding more nutrients is a good thing just does not hold up.

Mr. BARTLETT. You are right. But I do not think having a bit more CO2 in the oceans that anybody will argue but what that would increase the phytoplankton there at the beginning of the food chain.

How long does this equilibrium process take? When you are putting CO_2 in the air, how long does it take to equilibrate?

Mr. MICHAELS. There has been a statement made over here about 120-year lifetime of CO_2 in the atmosphere.

Mr. SINGER. That is not quite correct. The leading researcher in the field is Professor Sarmiento at Princeton, who is the lead author of one of the IPCC chapters. He gives 30 years as the mean life for CO_2 in the atmosphere, which is a factor of 4 less than some of these numbers that have been bandied about here today.

Now, a very small amount of the CO_2 will remain for long periods of time, but half of it would disappear after 30 years.

Mr. BARTLETT. Half-life equilibration is something like 30 years?

Mr. LASHOF. No. That is the problem. It is not a simple exponential decay because there are several time scales involved and that is why you hear different numbers. It takes a while.

Mr. BARTLETT. So if we were to level off CO_2 production now, one would expect it would go down in the atmosphere as the equilibrium with the oceans took place.

Mr. LASHOF. No.

Mr. BARTLETT. That is the only conclusion I can draw from what you just told me.

Chairman TALENT. Is it scientifically possible that it might go down?

Mr. SINGER. Yes, it will eventually disappear because the ocean has to absorb the excess CO_2.

Chairman TALENT. That would be very bad for the ecosystem, correct? So applying the precautionary principle, Dr. Lashof, aren't we not to try to increase the amount of CO_2 here?

Mr. LASHOF. No.

Chairman TALENT. Why do you laugh when I say that and you are very serious when you say it?

Mr. LASHOF. I am sorry, Mr. Chairman, but it has been a long morning.

Chairman TALENT. It has been.

Mr. MICHAELS. It is not morning, Dan.

Mr. LASHOF. The question was, if you level off emissions at current levels, what happens to CO_2? All carbon cycle models that I have seen suggest that carbon dioxide concentrations continue to increase for centuries if you do that. The reason is we are so far out of equilibrium now that if you just maintain emissions at current levels you stay out of equilibrium for several centuries and concentrations rise very dramatically before you can establish an equilibrium.

Now, if you cut emissions to zero, then I would, first, agree with the Chairman that that would have a drastic effect on the economy, if you did it quickly; second, CO_2 concentrations would fall. They would fall very slowly.

Mr. BARTLETT. You are arguing now for very long equilibrium periods.

There is a lot of movement of molecules in the air. There is a lot of movement in the ocean with waves and so forth. I would be surprised if one could postulate scores of years.

Mr. LASHOF. The reason for that is the long-term equilibration is determined by the overall mixing time of the ocean. The ocean is not a well-mixed reservoir and the overall mixing time of the ocean is more on the order of 500 years. There is a top layer of the ocean, about the top 75 meters, that is well-mixed by winds. But below that the ocean is highly stratified.

So you have part of the CO2 that equilibrates quickly with the upper part of the ocean. But then for the full equilibration, which is where you get most of the removal, the full depth of the ocean, the other 4,000 meters, requires centuries.

Mr. MICHAELS. Mr. Bartlett, I would like to point out that if we read the record of today's proceedings we will see that my two colleagues on the right have said that the residence time of carbon dioxide in the atmosphere is 100 years, 120 years, and 30 to 120 years.

Mr. BARTLETT. We just do not know, is that what we are saying?

Mr. LASHOF. It is not a well-defined question is the problem. There is not a single residence time. It is not exponential decay.

Mr. BARTLETT. But do we still have a ways to go? There still is a sink out there that will pick up the CO2? That is a correct statement?

Mr. LASHOF. That is correct. That is fully accounted for in the projection that we have been talking about.

Mr. WATSON. Indeed, we expect that roughly two gigatons of the anthropogenic carbon, human carbon, goes into the oceans every year. The energy emissions are about 6 billion tons per year. We expect at least 2 billion tons of that is absorbed by the ocean and that is fully reflected in the models.

Chairman TALENT. I am going to, Dr. Watson, since you gave up a trip to come here, I believe I am going to let you have the last word. I want to thank everybody for being here this morning and for your patience with the Committee and our questions, which often may have been perhaps too passionate. But it is a very important subject. On that I think we all agree.

I thank the witnesses for being here and adjourn the hearing.

[Whereupon, at 2:12 p.m., the Committee was adjourned, subject to the call of the Chair.]

APPENDIX

Chairman Jim Talent

TESTIMONY

Full Committee Hearing
Kyoto Protocol: The Undermining of American Prosperity?
The Science
July 29, 1998

Opening Statement of James M. Talent
Chairman
House Small Business Committee

Today, the Committee on Small Business will be holding its third hearing on the impact of the Kyoto Protocol, in which the United States agreed to reduce its production of greenhouse gases to below 1990 levels. Our previous hearings established that this is equivalent to raising the cost of energy to the American economy by 30-50%. For Americans this would mean:

1. Gasoline prices would increase by 65 cents per gallon;
2. Natural gas prices for industry would increase by 90%; and
3. Gross Domestic Product would decline by 2.5%.

The proponents of the Kyoto Protocol justify this effect on the American economy by the need to prevent what they describe as catastrophic warming of the Earth's temperature.

I called this hearing with scientific witnesses to determine whether and to what extend the Earth is actually warming. My research into the subject suggests that the catastrophic global warming predictions of Kyoto proponents are based primarily on

1. The pretense of a consensus within the scientific community, when in fact no such consensus exists;
2. Highly selective and misleading use of data; and
3. Utter failure to confront the body of evidence suggesting that the Earth's warming trend is natural and beneficial.

We have a number of distinguished scientists here, several of whom will attempt to disabuse me of these views. The Committee is grateful for the time of these witnesses. I will introduce them later, and now recognize the ranking member, the distinguished gentlelady from New York.

50-206 98 - 3

Verbal Remarks: John R. Christy, 27 July 1998
House Committee on Small Business

Thank you, Mr. Chairman, and members of the Committee on Small Business for inviting me to provide some information on the global climate. Please note that I will be referring to figures which you should have attached to your copies of my remarks.

Twenty years is nothing in terms of the Earth's climate, but twenty years is everything in terms of the **science** of climate as shown in Figure 1.

In the late-1970's, the climate stories were filled with the potential disasters of the coming Ice Age. I lived in South Dakota during the coldest winter of that decade and we can only hope those ice-age forecasts of the 70's will never be realized.

Ten years ago, the potential catastrophe of human-induced global warming began receiving a lot of attention, thanks in part to a couple of warm (but not record breaking) summers in the eastern US. The predictions were horrifying ... rapid temperature rises, coastal flooding, massive hurricanes and so on.

Twenty years now have passed since the 1970's and we are now almost exclusively discussing global warming brought to the forefront ten years ago. Twenty years makes **all** the difference in the science of climate.

When global warming became a popular issue ten years ago, I and others were concerned with the lack of proper data to describe the Earth system and a lack of perspective in which to judge the quasi-extreme events occurring at that time ... events so often connected with human-induced climate change. In 1989, NASA scientist Dr. Roy Spencer and I set out to create a satellite dataset of truly global observations of atmospheric temperature. These data would not be plagued as the traditional surface data are by changing locations or dependence on transient shipping or lack of coverage for large areas. We did not know what this dataset would show, our goal was precision ... to provide scientists with excellent, truly global temperature data.

An added incentive for us was that two of the atmospheric layers we could observe happened to be two layers for which climate models indicated the largest and most rapid responses to human-induced climate change would occur - the lower troposphere (sfc to 20,000 ft) warming faster even than the surface, and the stratosphere which would cool. In the data presented here, all known artificial effects, such as orbit decay and orbit drift, have been removed.

Figure 2 shows the tropospheric temperature since 1979 has little trend, either up or down, while the stratosphere shows a decline, being heavily influenced by volcanic events and ozone depletion.

Were these data accurate? Balloons also measure the atmospheric temperature at various places around the world. I and others have compared the satellite data with many estimates of regional and global temperature, with one example shown in Figure 3 produced by the U.K. Meteorological Office using about 400 balloon stations. The comparison shows the correlation is high and there is no warming in either dataset for 1979-97. The satellites were truly providing the precision for which we had hoped. This is that same period in which the **science** of climate shifted from thinking mostly about global cooling to thinking mostly about global warming.

If we look at the comparison since 1979 between the surface and the atmosphere in Figure 4, the temperatures seem to be going in somewhat different directions. Now, we have only twenty years of satellite data – too short a period to judge long-term climate variations. However, these data tell us that **in** this layer and **during** this period when CO2 warming should be most pronounced, we see no trend. We also see something unpredicted by climate models – a difference in trend between surface temperatures and the deep atmosphere. Could the consensus about global warming reached by these same climate models be misrepresentative of the next century? Twenty years of truly global data are important to the science of climate.

The recent fixation on extreme events as indicators of climate change is misleading because we know very little about the rates of their occurrence and we are able to publicize even marginal extremes to fantastic proportions. Perspective is often lost in the media, though fortunately we have considerable historical information archived at the National Climatic Data Center (NCDC).

For example, last month, a government-sponsored press conference informed you that five states, MD, CT, MA, VT and NJ experienced one record monthly high temperature at some point in the period January through May of this year. Human-induced warming was implicated in that announcement. Now, these five states would fit into the state of Kansas TWICE – not a very geographically representative sample. However, let me put a different spin on record high temperatures, provided through NCDC, and you will see my point. If one divides the recorded climate history of the entire U.S. into halves at 1940 and looks at all-time record high temperatures, one would see that 4 states had equal highs in both halves. Of the remaining 46 states, 35 recorded their all-time record high temperature prior to 1940 and only 11 have since 1940. Or, take the Midwest where newly digitized information from early years now provides a better look at extremes. For the ten hottest 3-day periods (i.e. killer heat waves) in the Midwest since 1896, only two occurred after 1940, and they placed 7th and 8th.[1] Do these results prove cooling? No. They prove that the spin one places on extreme events can be very misleading.

---

[1] **K. Kunkel, Midwestern Regional Climate Center**

The recent attention given to the heat in the South is another example. An A.P. story declared June 1998 the hottest in Huntsville, Alabama's history. I was surprised that our Office of the State Climatologist hadn't been asked about this record which I knew to be misleading. So, I did some checking and found at least 6 other years between 1914 and 1953 in which June was hotter than in 1998. I sent the information to the local paper which eventually published it. The problem was that a local weather service forecasting office, which reporters typically call, did not have access to records before 1958, and thus could not have known about the earlier, hotter years.

Again, the Florida fires and drought were highlighted every evening on the news. Was this drought remarkable? The answer from climate records at NCDC is no. In Figure 5, I show the percentage of months in which Northern Florida experienced drought conditions decade by decade. In this century you can see that in only one other decade – the 1940's – were there fewer drought months than the 1990's.

Yes, it was hot in the South, but this was due to a weather pattern that placed warm air in the South, and cool air in the West. To a climatologist it was equally remarkable that towns in the High Plains recorded their latest snowfall ever in June 1998. Or that Fresno, California, my hometown, for the first time in its history failed to reach 100°F by the middle of July. In fact, for the nation as a whole, the temperature in June 1998 was **below** normal.

To see the real picture of extreme weather variations given by droughts and floods from 1895, turn to Figure 6. In this chart, again kindly provided by NCDC, is the percentage of area in the lower 48 states which in any given month experienced extreme conditions, dry or wet. The results show there is no significant long term trend in either condition. In fact, we have not had a widespread drought in several years, and looking at the chart, we should expect one in the near future, based solely on statistical probability. My point is this: arriving at conclusions about human-induced climate change based on selective extremes is faulty and misleading. In the United States, the evidence presented here suggests human-induced climate change is not (yet) discernable.

The satellite and balloon data show that significant global warming is not now occurring, though the simple theory and laboratory experiments of greenhouse warming are indisputable. The detection of human effects on the Earth's climate has not been convincingly proven because (1) the variations we have observed to date are not outside of the natural variations of the system and (2) the Earth system is very complex and probably has ways to expel heat energy which are not properly accounted for in climate models. The stratospheric temperatures suggest something is going on, but separating the massive effects of the volcanoes and ozone depletion (effects unrelated to fossil fuel emissions) is not easily done.

I believe some minor warming due to increasing greenhouse gasses will occur, though untangling that component from the natural variability will be extremely difficult. Even drastic actions at this point will have a miniscule effect on whatever the present course of the climate is. We do know that a CO_2-enriched **atmosphere** is beneficial to all our carbon-based ecosystems. However, at this point, we simply do not know whether the CO_2-enriched **climate** of the next century will be made more difficult or more benign or even noticeably different for our country or for the world. The research is really just beginning.

I will close with a personal observation. Some activists cultivate the notion that they are taking the High Moral Ground by putting global warming on the same "Save the Earth" platform as preserving wildlife, saving the rain forests or cleaning the Earth's rivers and oceans.

Serving as a missionary in Africa in the early 1970s gives me a different perspective on proposals to combat global warming. The artificial rise in fuel and energy costs created by the Arab oil embargo was a crushing blow to many, but in particular to the world's poorest peoples. Many people in Africa suffered or died when gasoline prices soared beyond their meager ability to pay, halting transportation and other energy-related activities that preserved both life and property.

In addition, many of the life-saving activities supported through the benevolence of wealthy countries, such as ours, were curtailed due to rising costs and slumping economies.

Disrupting the lives of those whose existence is too often literally hanging by a thread causes the kind of suffering that the average policy maker or activist never sees. I have seen it. It is real and it is devastating.

That is why my own view, based on personal experience as both an atmospheric scientist and a missionary, leads me to a different High Moral Ground: When it comes to global warming, save the Earth but not before you reduce the scientific uncertainties and not by sacrificing the poor. Thank you.

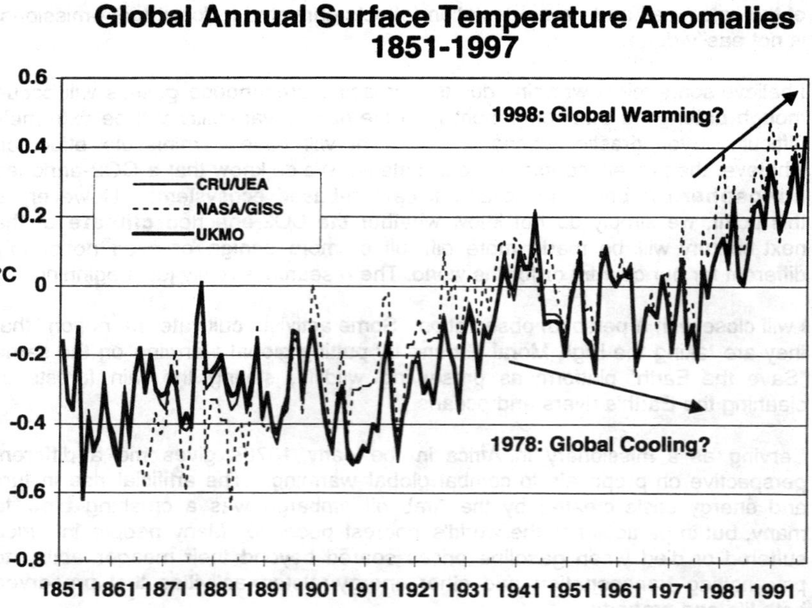

Figure 1

Surface Data Sources:
Climate Research Unit, Univ. E. Anglia, P.D. Jones
Goddard Institute for Space Studies, NASA, J. Hansen
U.K. Meteorological Office, D. Parker

Monthly Global Lower Troposphere Temperatures 1979-98 (Satellite MSU, Spencer/Christy, T2LT.d01)

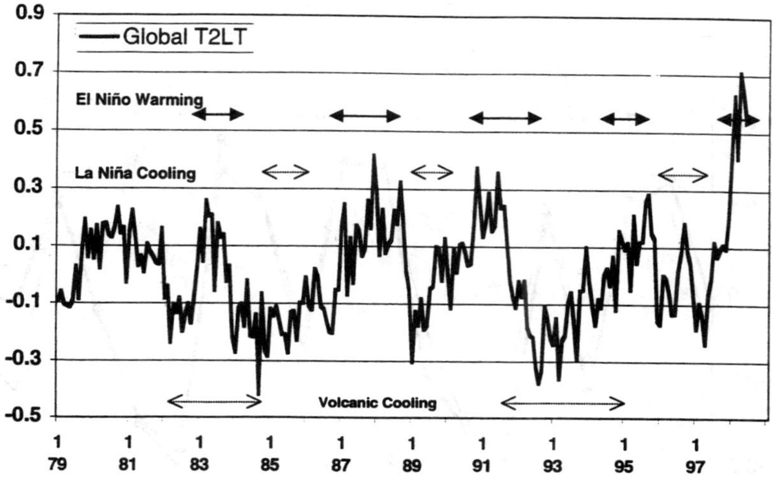

Figure 2

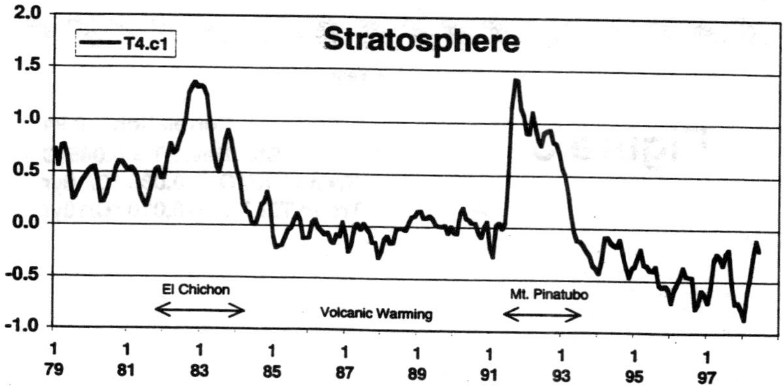

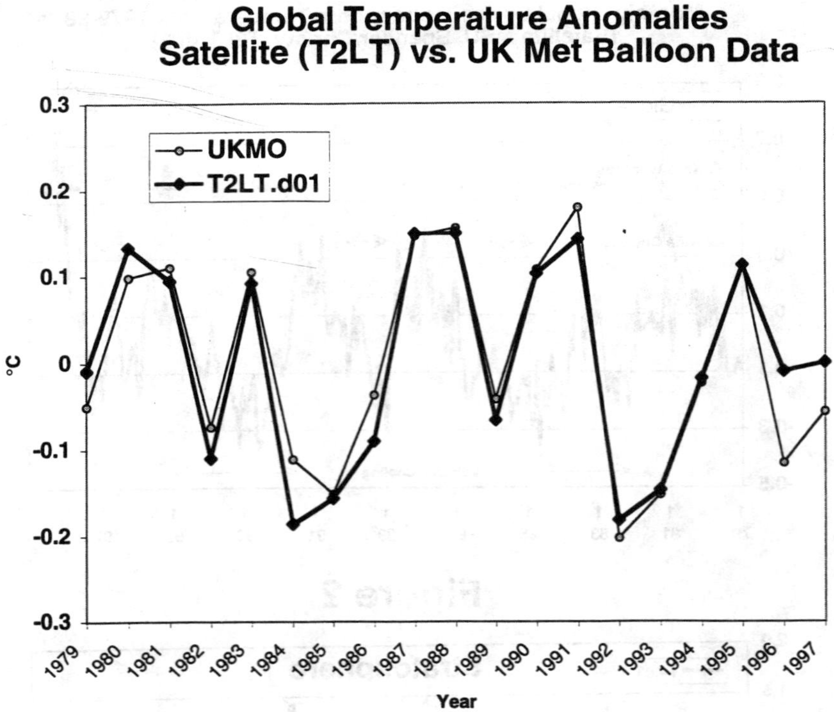

Figure 3

Correlation = 0.94
Std. Dev. ΔT = 0.045°C
Trend UKMO = -0.035 °C/10yr
Trend T2LT.d = -0.010 °C/10yr

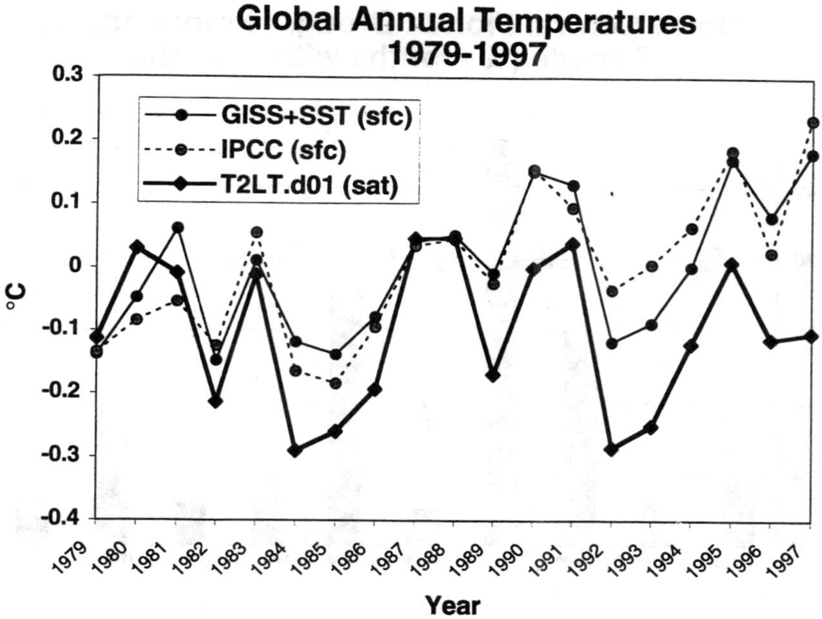

Figure 4

Trends °C/Decade

GISS+SST Global +0.11
IPCC Global +0.15
Satellite Global -0.01

Std. Dev.
ΔT IPCC v. GISS = 0.053 °C

Correlations

IPCC v. GISS+SST = 0.88
IPCC v. Satellite = 0.55
GISS+SST v. Satellite = 0.71
IPCC v. UKMO = 0.57
GISS+SST v. UKMO = 0.73

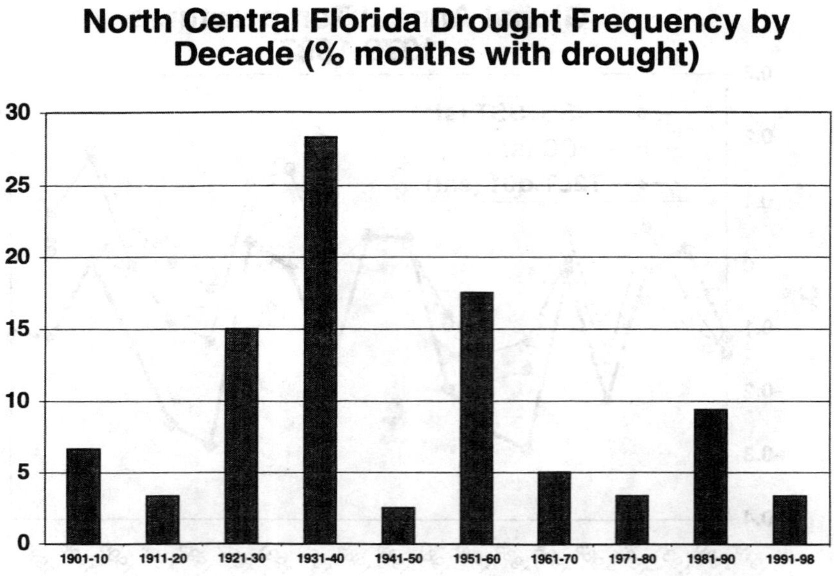
North Central Florida Drought Frequency by Decade (% months with drought)
30
25
20
15
10
5
0
1901-10
1911-20
1921-30
1931-40
1941-50
1951-60
1961-70
1971-80
1981-90
1991-98

Figure 5

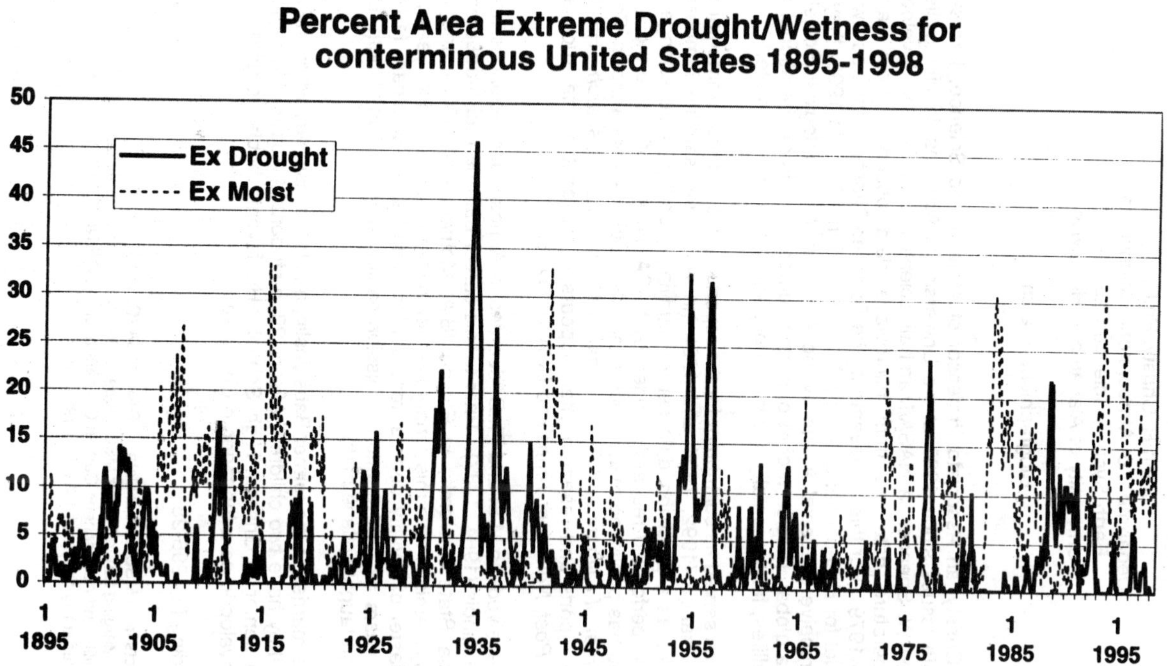

Figure 6

Drought trend -0.12% per decade
Moist Trend +0.15% per decade

John R. Christy, Ph.D.
Department of Atmospheric Science and
Earth System Science Laboratory
University of Alabama in Huntsville

Biographical Sketch

Dr. John R. Christy, an Associate Professor of Atmospheric Science, has studied global climate issues since 1987 at the University of Alabama in Huntsville. In 1989 Dr. Roy W. Spencer, a NASA/Marshall scientist, and Christy developed a global temperature data set from microwave data observed from satellites beginning in 1979. For this achievement, the Spencer-Christy team was awarded NASA's Medal for Exceptional Scientific Achievement in 1991. In 1996, they were selected to receive a Special Award by the American Meteorological Society "for developing a global, precise record of earth's temperature from operational polar-orbiting satellites, fundamentally advancing our ability to monitor climate."

Dr. Christy has served as contributor for the reports by the Intergovernmental Panel on Climate Change (1992, 1994 and 1995) in which the satellite temperatures were included as a precision data set for studying global climate change since 1979. He has performed research funded by NASA, NOAA, DOE and the State of Alabama and has published many articles including studies appearing in *Science, Nature, Journal of Climate* and *The Journal of Geophysical Research.* His global temperature reports and research often appears in the print media such as the *Washington Post*, *New York Times* and *USA Today.*

Dr. Christy received the M.S. and Ph.D. degrees in Atmospheric Sciences from the University of Illinois (1984, 1987). Prior to this career path he had graduated from the California State University in Fresno (B.A. Mathematics, 1973) and taught Physics and Chemistry as a missionary teacher in Nyeri, Kenya for two years. After earning a Master of Divinity degree from Golden Gate Baptist Seminary (1978) he served four years as a bivocational mission-pastor in Vermillion, South Dakota where he also taught college math.

Dr. Christy is married to the former Babs Joslin, a fellow missionary whom he met in Kenya, and they have two children, Alison, 20, a student at Auburn University and Brian, 17, a senior at Grissom High School. His favorite hobby is gold-panning which he developed as a teenager in California.

Federal funding in 1996-98

Climate Products from MSU Global Data - NOAA/NASA (Co-I) 1997-2000	$190k
Optimizing the Merging Strategies for MSU Global Data - NASA HQ/ Pathfinder (P-I) 1995-98	$125k
Climate Variability and Atmospheric Carbon Concentrations -DOE/NIGEC (P-I) 1995-98	$270k
Atmospheric Variability Studies - TVA (P-I) 1994-98	$ 750k

NATURAL RESOURCES DEFENSE COUNCIL

Statement of

Daniel A. Lashof, Ph.D.
Senior Scientist
Natural Resources Defense Council

Before the

Committee on Small Business
United States House of Representatives

July 27, 1998

Thank you Mr. Chairman, I appreciate the opportunity to appear before the Committee on the important matter of global warming. My name is Daniel Lashof, and I am a senior scientist with the Natural Resources Defense Council (NRDC). I have been studying global climate change for more than a decade and have contributed a number of peer-reviewed papers to the scientific literature. I hold a Ph.D. from the University of California and a B.A. from Harvard.

NRDC is a nationwide, non-profit organization with over 350,000 members, dedicated to protecting human health and the environment.

Mr. Chairman, it is an indisputable fact that atmospheric concentrations of greenhouse gases are climbing and that this rising trend can be attributed to human activities. The most detailed scientific assessment of an environmental problem ever conducted—the Second Assessment Report of the Intergovernmental Panel on Climate Change—concluded that global warming is already upon us, and that, unless emissions of greenhouse gases are reduced significantly, future impacts on human health and the environment will be severe, and are likely to include:

00 New York Ave., N.W.
Suite 400
Washington, DC 20005
202 289-6868
Fax 202 289-1060
www.nrdc.org

71 Stevenson Street
Suite 1825
San Francisco, CA 94105
415 777-0220
Fax 415 495-5996

6310 San Vicente Boulevard
Suite 250
Los Angeles, CA 90048
213 934-6900
Fax 213 934-1210

40 West 20th Street
New York, NY 10011
212 727-2700
Fax 212 727-1773

- **More frequent and intense heat waves.** As average temperatures rise over the coming century, the frequency of severe heat waves will increase dramatically, which is expected to result in an increase in heat-related deaths.
    - ⇒ Each month so far this year has set a new all-time record for global near-surface temperatures. June of 1998 exceeded the previous record warm June of 1994 by nearly 0.4 degree F (see Exhibit 1).
    - ⇒ The current heat wave, which has claimed more than 100 lives in the United States this year, as well as the heat wave that killed over 400 people in Chicago in 1995, are example of the kind of event that is expected to occur with increasing regularity. According to NOAA:
        - * Brownsville, Texas recorded 17 days this June that had minimum temperatures at or above 80F. This has a probability of recurrence in a stable climate of once in 1000 years. By the middle of the next Century this would be a 1 in three year event presuming greenhouse gases continued to increase at the present rate.
        - * In Melbourne, Florida there were 24 days with temperatures above 95F this past June. This is an unprecedented event with a probability of recurrence of much less than 1 in 1000 years in a stable climate.

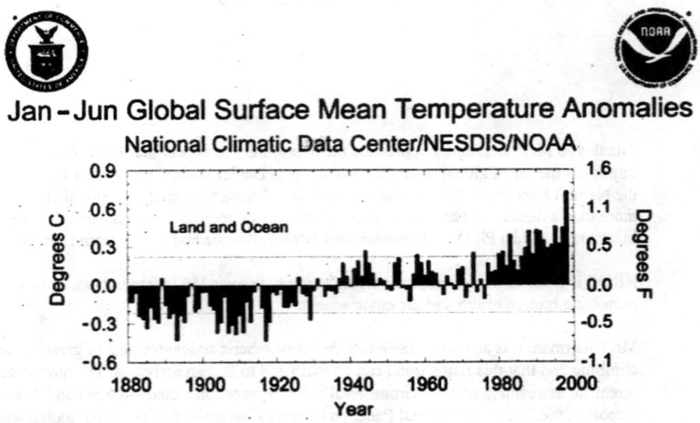

Exhibit 1

- **Increased spread of infectious diseases.** Infectious diseases that are carried by insects and other pests are highly sensitive to environmental conditions. Higher temperatures will increase the range of malaria-carrying mosquitoes and make the mosquitoes more infectious. Global warming could cause as many as one-million additional deaths from malaria each year during the middle of the next century. Other sensitive diseases include hantavirus and dengue fever, which has already broken out in Texas.

- **Flooding along our coasts and river basins.** Warmer temperatures cause sea levels to rise and allow the atmosphere to hold more moisture. In addition to coastal flooding, inland areas can be affected by earlier and more rapid snow melt and more intense storms, leading to greater rates of flooding, such as the ones witnessed last year in North Dakota and Minnesota.

- **Extreme droughts and more severe storms in our heartland.** More of our rainfall is now coming in intense downpours rather than steady drizzles. This coupled with higher rates of evaporation in a warmer world is expected to intensify droughts and storm damage in America's breadbasket.
  - ⇒ The current drought in the Southeast United States is a clear example of this kind of event. Last winter was very wet, with soil moisture levels in the Southeast well above normal. Yet severe drought conditions developed very rapidly this Spring due to intense evaporation when the normal rains failed, as shown in Exhibit 2.
  - ⇒ The wet winter led to extensive growth of underbrush which fueled the fires when drought conditions developed. Wildfires in Florida consumed almost half of a million acres this June. More than 140,000 acres have burned in Texas. Meanwhile, drought-induced fires have burned more than 3 million acres in Mexico and Central America.
  - ⇒ Texas now estimates over $1.5 billion in agricultural losses, Oklahoma about $2.0 billion, Florida about $175 million, and Georgia over $400 million as a result of the drought.

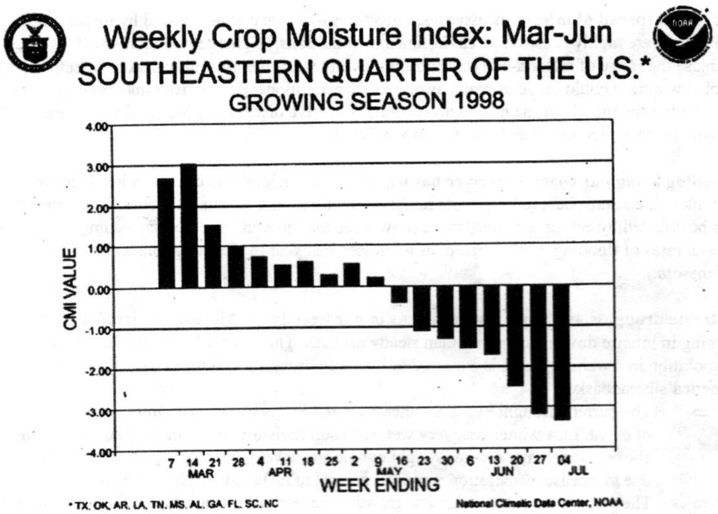

Exhibit 2

The climate system is complex, and there are remaining uncertainties, particularly about the details of climate change at the regional and local level. While it is impossible to unequivocally demonstrate a link between global warming and any specific event, the examples I have given do demonstrate that global warming can have severe consequences. It is abundantly clear that we are facing a real and serious threat.

As we discuss what we still don't know, it is essential to remember that uncertainty is a double-edged sword. We cannot rule out the possibility that climate change will be less severe than expected in the mainstream scenario outlined above, but it is just as likely that things will turn out to be worse than the mainstream forecast. For example, in a recent journal article I reviewed feedback mechanisms that have the potential to substantially exacerbate global warming beyond levels considered in the mainstream projections. Another recent article reviews the risk that sea levels could rise by as much as 20 feet over the next several centuries, rather than the mainstream forecast of about five feet. These articles are attached to my testimony for the record.

The Defense Department would never base its response to military threats on the views of a handful of doubters or on the hope that the world will turn out to be benign. We should apply the same precautionary principle to global warming policy.

As a party to the United Nations Framework Convention on Climate Change, signed by President Bush and ratified by the Senate in 1992, the United States is officially committed to this precautionary approach. The 1992 Convention sets the objective of preventing dangerous anthropogenic interference with the climate system. Achieving this objective requires a substantial and progressive reduction in global greenhouse gas emissions beginning without delay.

Thank you, Mr. Chairman.

COMPETITIVE ENTERPRISE INSTITUTE

Why Kyoto Is Not an Insurance Policy

Marlo Lewis, Jr.
Vice President for Policy and Coalitions,
Competitive Enterprise Institute

House Small Business Committee
July 27, 1998

Thank you, Mr. Chairman, for inviting me to present testimony on behalf of the Competitive Enterprise Institute, a non-profit research and advocacy group committed to advancing the principles of free enterprise and limited government.

What I'd like to do in the next several minutes is address what I take to be the Clinton Administration's strongest argument in favor of an international climate treaty. This is the argument that, "if we do it smart," Kyoto will provide low-cost planet insurance for ourselves and future generations.

In candid moments, Administration spokespersons will admit that the theory of catastrophic warming has not been validated by experimental or empirical evidence. They'll concede that scientists know too little about the underlying physics, that computer models are too slow, and that the evidence is too conflicted, to permit a genuine resolution of the global warming debate. In other words, they'll admit, at least privately, that the science supporting the Kyoto Protocol isn't really clear, compelling, or "settled." But they don't see this as a great liability. Indeed, in their view, our very ignorance about the extent of human influence on the climate system is reason enough to justify an enterprise like the Kyoto Protocol.

I. Precautionary Deception

The Administration's trump-card argument in the global warming debate is not any testable scientific hypothesis but something called the Precautionary Principle. This is the proposition that lack of scientific certainty should not become an excuse for inaction where there are threats of serious or irreversible harm to health, safety, or biodiversity.[1] The precautionary case for Kyoto goes as follows. The earth *may* be warming; industrial activity *may* be the cause; and the effects of such warming *may* be catastrophic. Furthermore, a global program to reduce greenhouse gas emissions *may* be feasible and effective in averting or mitigating harmful climate change.

Mankind, through its combustion of fossil fuels, has been running a gigantic, uncontrolled experiment on the climate system. Since this experiment is potentially life threatening, we should start applying some controls. Curbing energy use to reduce emissions could prove costly, but what is money compared to the lives that might otherwise be lost? Kyoto or some similar treaty is, thus, the only responsible option. The alternative is to throw caution to the winds and gamble with "the only planet we have."

This argument is rhetorically powerful because it sounds so much like familiar maxims of common sense – look before you leap, err on the side of caution, better safe than sorry. In fact, as will soon become clear, we can with equal legitimacy invoke precautionary considerations to oppose the Kyoto Protocol. My testimony has two parts. First, I'll show that the Precautionary Principle supplies no real guidance for choosing between policy alternatives. Then I'll offer specific reasons why Kyoto is not an insurance policy.

The Precautionary Principle Is Self-Contradictory

The fatal flaw in the precautionary case for Kyoto – as in environmental advocacy generally – is its complete one-sidedness. Environmentalists demand assurances of no harm only with respect to actions that government might regulate, never with respect to government regulation itself. But government intervention frequently boomerangs, creating the very risks that precautionists deem intolerable.

Examples abound. Federal fuel economy mandates force auto makers to produce smaller, slighter, less crash-resistant cars, causing 2,000 to 4,000 highway deaths per year. FDA regulations delay the availability of life-saving therapies, killing tens of thousands over the past

---

[1] Words to this effect occur in both the 1992 Rio Treaty and the Biodiversity Convention. See James Cameron, "The Status of the Precautionary Principle in International Law," in Timothy O'Riordan & James Cameron, eds., *Interpreting the Precautionary Principle* (London: Earthscan Publications, 1994), p. 269.

two decades. Banning DDT revived malaria epidemics in the Third World, afflicting 2.5 million people in Sri Lanka alone.

Frank Cross at the University of Texas at Austin notes that over-regulation can kill just by misdirecting resources and destroying wealth. The resources available to protect public health and safety are limited. Regulatory schemes that divert attention, effort, and money from major threats to minor risks make us less safe. For example, the millions of dollars local governments waste on gold-plated Superfund cleanups cannot be used to improve police and fire protection.

Even more important is the fact that, for individuals as well as nations, wealthier is healthier and richer is safer. Precautionists ignore the obvious connection between livelihoods, living standards, and lives. Wealth is the single most important factor affecting health and longevity. For example, a recent study published in *JAMA* finds that the death rate of America's poor is three times that of the general population. Only 13% of the difference can be explained by risky lifestyle choices such as overeating, excessive drinking, or smoking. What's chiefly causing the higher death rate is poverty itself. Poverty is stressful and highly correlated with several well-known disadvantages – unsafe neighborhoods, dangerous jobs, unhygienic living conditions, inadequate medical care. Even in comparatively affluent countries like the United States, studies indicate that every $5 million to $10 million drop in economic output translates into one statistical death.

So here is a precautionary argument *against* the Kyoto Protocol. Stabilizing greenhouse gases at levels low enough to cool the planet *may* require drastic reductions in energy use (actually, I think this is a near certainty, but I'll say "may" to preserve the parallelism). An energy-constrained world *may* be a poorer world (again, this is a virtual certainty).. A poorer world *may* be a world with more starving people (indeed, how could it not be?). True, I do not have scientific proof that Kyoto would condemn millions to poverty, starvation, and misery. But the Precautionary Principle says that we should not let the absence of scientific certainty become an excuse for inaction. It also says we should not permit far-reaching innovations until they are proved to be safe. Therefore, until the Kyoto Protocol is proved to be safe, we should oppose it.

Kyoto partisans sometimes compare mankind to a motorist who suddenly finds himself driving down an unlit road on a foggy night. Until we know what lies ahead, shouldn't we take our collective foot off the accelerator? That depends on whether the analogy is apt. Perhaps mankind is more like a pilot flying into dark clouds. If the pilot reduces air speed too much, he may lose altitude and crash.

Only a small fraction of mankind enjoys the living standards we Americans do. Affordable energy is one of the key foundations of our prosperity. As Mark Mills of Mills-McCarthy Associates has shown, there is an amazing correlation throughout the world between per capita energy consumption and per capita income. Where per capita energy consumption is high, per capita income is also high; and where per capita energy consumption is low, per capita income is also low. Thus, if we are to rescue mankind from the perils of poverty, we must dramatically increase global energy consumption – we must push down on the accelerator. In countries like India, there are people so poor that they burn dung for fuel and poison themselves breathing the polluted air of their own homes. The world isn't rich enough to justify any effort to slow down the airplane. Rather, what's needed are better communications systems, improved early warning systems, a faster, sturdier, more resilient aircraft. We are more likely to develop such assets in a free, energy-abundant economy than in an over-regulated, energy-poor economy.

For far too long, environmentalists have gotten away with precautionary deception. They say we should not permit new products, technologies, or industrial processes until those innovations are proved to be safe. Yet they are willing to launch new regulatory schemes without a thought about their potentially lethal effects. In the global warming debate, they admonish us not to gamble with the only planet we have. Yet they are more than willing to gamble with the only economy we have. They cannot logically have it both ways. They cannot consistently say that "safety first" trumps all other considerations in the realm of private action but has no application in the realm of governmental action.

Now, I am not for a moment suggesting that the Precautionary Principle would be beneficial were it applied evenhandedly to economic and regulatory activity alike. Acting as if "Safety First!" were a categorical imperative rather than a mere rule of thumb is a recipe for paralysis and stagnation – perhaps the riskiest condition of all. My point, rather, is that the Precautionary Principle is neither a scientific nor an ethical precept. It is a rhetorical weapon. Its purpose is to inflate the importance of certain kinds of risks – those associated with economic endeavor – and conceal the importance of other kinds of risks – those arising from the exercise of political power.

An honest presentation of the subject begins by acknowledging that there are risks on both sides of the ledger. There are risks of under-regulation but also of over-regulation. Risks of climate change but also of climate change policy. Do the risks of climate change outweigh those of climate change policy? Or do we have more to fear from Kyoto than from global warming itself? Those are the questions policy makers should be debating.

II. Kyoto Is Not an Insurance Policy

The Administration claims that Kyoto is an insurance policy, and insurance by definition is supposed to make us safer. But is Kyoto the real thing, or is it as phony as the Precautionary Principle?

Whether or not a proposed insurance policy makes sense depends chiefly on three things: the probability and size of the potential losses to be insured against, the cost of the premiums, and the extent of the coverage. Earthquakes are very real and sometimes catastrophic events. But they seldom occur in Washington, D.C. – so seldom that even low-cost earthquake insurance would be a bad investment for homeowners in the District. Fires do destroy homes in this area, and it would be unwise not to have fire insurance. But how much should one be willing to pay? A policy so expensive that it effectively prevents the homeowner from purchasing health care, fixing the brakes on his car, or saving for his retirement, would probably make him less safe. Finally, an insurance policy worthy of the name should help make the insured whole after misfortune has struck. Or, if it can't do that, the policy should at least reduce the odds of the bad thing happening. If the policy can't either reduce the likelihood or cushion the impact of the insured-against event, it's not worth beans.

These simple reflections suggest that we ask three questions of the Kyoto insurance salesmen: What is the likelihood of a human-induced global warming catastrophe? How much will the premium payments cost? How much protection will Kyoto insurance deliver?

How Likely Is Catastrophic Warming?

There are several reasons why we shouldn't worry about global warming.

- The computer models on which the Rio Treaty is based overestimate by 200%-300% the amount of warming recorded over the past century. If the models can't retrodict the past, why should we suppose they can accurately forecast the future? The fact that the earth has warmed only 1 degree F over the past 150 years suggests that the climate system is less sensitive to greenhouse "forcing" than the models assume.

- Greenhouse theory says the troposphere should be warming faster than the surface, and that the largest warming signal should have occurred in recent decades. Yet the satellite and weather balloon records of the past 20 years show no warming trend in the troposphere.

- The models assume that the slight increase in heat energy from man-made carbon dioxide will increase the moisture content of the upper troposphere, clogging the earth's heat exhaust vents. Empirical observations from satellites fail to confirm this.

- Most of the warming in this century occurred before most of the buildup in man-made greenhouse gases. All of that warming is within the natural range of climate variability.

- When the planet was warmer than today (e.g., the Medieval Warming of 1000 AD to 1200 AD), mankind flourished. When the planet was cooler (e.g., the Little Ice Age of 1300 AD to 1850 AD), mankind suffered.

- In the Atlantic basin, the intensity of hurricanes and the number of hurricanes making landfall have been decreasing since the 1940s.

- The one indisputable thing we know about the buildup of atmospheric carbon dioxide is that it is helping trees, crops, and other plant life grow faster, stronger, and more profusely.

Such facts persuade me that the probability of catastrophic warming is low. Indeed, it is not clear global warming is something we should prevent, even if that were easy and cost little. Spending trillions to avoid better weather and a greener planet would make no sense at all.

How Much Will It Cost?

This brings us to the second question – how much will Kyoto insurance cost? The Administration says that if we "do it smart," Kyoto won't cost much at all – maybe one-tenth of 1% of GDP. Yet the Protocol would require America to reduce its emissions of greenhouse gases 31% below the level projected for 2010. In sheer physical terms, we'd have to cut emissions by 552 million metric tons per year during the initial compliance period – an amount equivalent to *all current emissions* from either the transportation sector, or the utilities sector, or industry. There is no feasible way to do that without putting our economy in the tank.

But, Administration officials claim, that is not the "smart" policy they're proposing. Through flexible market mechanisms like emissions trading, America will be able to pay others to reduce their emissions for a fraction of what we'd have to pay to reduce our own. In addition, federal R&D support for wind and solar power will further reduce compliance costs. This explanation is totally unpersuasive.

Emission trading has the potential to bring down costs, but by how much depends on the extent of the market, the number of countries participating. Developing countries are the largest potential source of inexpensive emissions credits. But the Kyoto Protocol exempts developing countries from binding emissions targets and, hence, from trading.

It is not even clear how much trading the U.S. would be permitted to do with other industrial countries. Article 17 states that emissions trading "shall be supplemental to domestic actions" undertaken to limit or reduce emissions. How much of our target we would be allowed to meet through "supplemental" trading rather than through "domestic actions" remains to be negotiated. The European Union has taken the position that no party should be allowed through trading to meet more than 50% of its obligations.

In recent issues of *World Climate Report*, Mark Mills demonstrates that if the EPA classifies carbon dioxide as a pollutant for purposes of capping emissions and facilitating trades, over a million small businesses become regulated stationary sources. Caught in EPA's regulatory net would be 28% of all schools and 25% of all health care facilities. To comply, these newly regulated entities would new staff to "install, identify, evaluate, and operate emissions monitoring equipment; some other people to undertake record-keeping and documentation control; yet another team to become expert in and monitor regulatory compliance; still others to consider and implement engineering solutions to the problem of complying with emissions reduction." Finally, legal staff would also be needed to cope with the liabilities arising under the new regulatory regime.

Mills estimates that these compliance activities will require one person-year of effort if done "in-house" and half a person-year if contracted out. The latter option, he conservatively calculates, will cost about $30,000. Since the threshold for coming under EPA's regulatory scrutiny is $8,000 in fossil fuel purchases, the cost of fossil fuel for a small business effectively jumps from $8,000 to $38,000 – a 300%-plus increase. Somewhat larger firms that spend $100,000 on fossil fuel will see their costs rise by "only" 30%. The total collective cost of these new regulatory burdens on small business could reach $100 billion annually.

What about those federal R&D investments in wind and solar power? Mills calculates that to fill just 30% of the projected growth in national electricity demand over the next two decades, wind and solar would have to increase their supply of electricity from about 10 million kilowatt-hours to 370 billion kilowatt hours. In other words, wind and solar will have to increase total output by 37,000 times! Asks Mills, "Does anyone really believe that any amount of taxpayer-funded Apollo program-style federal projects, tax incentives, outright grants, gifts, and similar silliness could lead to a 37,000-fold increase in solar/wind in just 20 years?"[2]

Clearly, Kyoto's premium payments will be higher than the Administration's estimates – perhaps staggeringly so.

How Much Protection Will It Provide?

The last question we must consider in assessing Kyoto as an insurance policy is how much protection it will provide. Here we should note that in popular parlance, "insurance" may refer either to a policy that reduces the likelihood of a bad thing happening, or to a policy that

---

[2] Mark Mills, "Energy Arithmetic," *World Climate Report*, July 13, 1998, p. 8.

reduces the impact of a bad thing once it has happened. Real insurance usually does some of both. For example, if one has a safe driving record, it is easier to qualify for auto insurance, or to get the best rates. Auto insurance thus encourages safe driving, reducing the likelihood of accidents. And if we have an accident, the policy will alleviate the financial impact.

Let's look at Kyoto under those two aspects. First, how much would Kyoto reduce the likelihood of global warming, assuming for the moment that the climate models are right? As Patrick Michaels demonstrates in his recent Cato Institute monograph, Kyoto would do next to nothing to cool the planet. Even if all nations – not just the industrial countries – cut their greenhouse gas emissions as much as Kyoto would require the U.S. to do, greenhouse gas concentrations would still rise over the next 50 years. By the year 2047, equivalent carbon dioxide concentrations would reach 602 parts per million rather than 655 parts per million. That 53 parts per million difference means that Kyoto would cool the planet by less than 0.2 degrees Celsius. "It is doubtful," observes Dr. Michaels, "that this change could even be extracted from ground-based temperature measurements, because of interannual climatic noise."[3]

Finally, would Kyoto be of any help in making us whole after harmful global warming has occurred? Here's where the insurance analogy completely breaks down. Kyoto can do nothing to put us back on our feet once we have suffered damages from destructive climatic change. Quite the contrary, whatever resources we apply under Kyoto to prevent climate change, we cannot use to adapt to climate change if and when it occurs. Kyoto is all premium and no coverage.

This last point leads to a question that should be at the heart of the global warming debate. *Which type of social insurance policy is likely to deliver the most protection?* Should we try to prevent global change by rationing energy and further politicizing economic and technological development? Or should we try to increase mankind's ability to adapt to change by reducing political barriers to enterprise, invention, and innovation?

Under the latter option – let's call it a Resiliency Strategy – policy makers would work to eliminate the political barriers (high taxes, burdensome regulations, wasteful subsidies) that impede economic growth and technological progress. Wealthier, more technologically advanced societies are better able to anticipate, withstand, and recover from extreme weather events and other natural disasters.

For example, both Florida and Bangladesh experience tropical storms. When a typhoon hits Bangladesh, tens of thousands of people may die. In contrast, when a hurricane strikes Florida, there may be tremendous property damage, but comparatively little loss of life. The

---

[3] Patrick Michaels, "The Consequences of Kyoto," *Cato Institute Policy Analysis*, No. 307, May 7, 1998, p.10.

50-206 98 - 4

reason is not that typhoons are worse than hurricanes but that Florida is wealthy, Bangladesh poor. In Florida, homes are built to withstand high-velocity winds. A developed infrastructure is in place to provide early warning, continuous communication, medical services, and emergency relief. Also, car ownership is widespread, allowing people to drive to safer locales before the storm hits. This suggests we should be exceedingly wary of policies that would stifle economic growth, distort technology development, or limit mobility.

A Resiliency Strategy is clearly the superior option given the uncertainties surrounding the global warming hypothesis. Many possible catastrophes may befall us in the next century – the outbreak of new viral plagues, increased tectonic plate activity, an errant meteor, nuclear terrorism, chemical or biological warfare, even global cooling. Making societies freer and wealthier is inherently desirable, and would better prepare us to deal with whatever shocks and surprises the future may hold. In sharp contrast to Kyoto, the Resiliency approach is all coverage and no premium.

The Kyoto Protocol would restrict our freedom, diminish our prosperity, and mobilize tremendous resources to fend off a threat that may prove to be non-existent or trivial compared to the age-old scourges of poverty, hunger, and disease. America does not need and cannot afford that kind of insurance.

Testimony of:

Patrick J. Michaels
Professor of Environmental Sciences
University of Virginia,
and Senior Fellow in Environmental Studies at Cato Institute

to:

Committee on Small Business
United States House of Representatives,

July 27, 1998

Thank you for soliciting my testimony on the science of climate change as it pertains to the Kyoto Protocol to the United Nations Framework Convention on Climate Change.

Nearly ten years ago, I first testified on climate change in the U.S. House of Representatives. At that time, I argued that forecasts of dramatic and deleterious global warming were likely to be in error because of the very modest climate changes that had been observed to that date. Further, it would eventually be recognized that this more moderate climate change would be inordinately directed into the winter and night, rather than the summer, and that this could be benign or even beneficial. I testified that the likely warming, based on the observed data, was between 1.0 and 1.5°C for doubling the natural carbon dioxide greenhouse effect.

The preceding paragraph was excerpted verbatim from my last testimony before this House, on November 6, 1997. Since that last testimony, new scientific advances have been published in the refereed literature that have now proven the validity of this position. The key findings include:

• Documentation that observed climate change is several *times* below the amount predicted by the climate models that served as the basis for the Framework Convention on Climate Change (Hansen et al., 1998),

• Documentation that observed changes are largely confined to winter in the very coldest continental airmasses of Siberia and northwestern North America (Balling et al., 1998),

• Documentation that the variation, or unpredictability, of regional temperatures has declined significantly on a global basis while there was is no change in precipitation (Michaels et al., 1998),

• Documentation that, in the United States, drought has decreased while flooding has not increased (Lins and Slack, 1997),

• Documentation that carbon dioxide is increasing in the atmosphere at a rate below the most conservative United Nations' scenarios, because it is being increasingly captured by growing vegetation (Hansen et al., 1998),

• Documentation that the second most important human greenhouse enhancer—methane—is not likely to increase appreciably in the next 100 years (Dlugokencky et al., 1998),

• Documentation that the direct warming effect of carbon dioxide was overestimated (Myhre et al., 1998), and

• Documentation that the Kyoto Protocol to the United Nations Framework Convention on Climate Change will have no discernable impact on global climate within any reasonable policy timeframe (Wigley, 1998).

In toto, these findings lead inescapably to the conclusion that the magnitude and the threat from global warming is greatly diminished. They should provoke a re-examination of the need for the United Nations Framework Convention on Climate Change, and the subsequent Kyoto Protocol.

Historical Background

Ten years ago, on June 23, 1988, NASA scientist James Hansen testified before the House of Representatives that there was a strong "cause and effect relationship" between observed temperatures and human emissions into the atmosphere. His testimony coincided with a very hot, dry period (much worse than the summer of 1998), and subsequent polls showed that, as a result of his testimony, the public believed that the 1988 drought was caused by human-induced global warming.

At that time, Hansen also produced a model of the future behavior of the globe's temperature, which he had turned into a video movie that was heavily shopped in Congress. That model was one of many similar calculations that were used in the First Scientific Assessment of the United Nations Intergovernmental Panel on Climate Change ("IPCC", 1990), which stated that "when the latest atmospheric models are run with the present concentrations of greenhouse gases, their simulation of climate is generally realistic on large scales."

That model predicted that global temperature between 1988 and 1997 would rise by 0.45°C (Figure 1). Figure 2 compares this to the observed temperature changes from three independent sources. Ground-based temperatures from the IPCC show a rise of 0.11°C, or more than four times less than Hansen predicted. Lower atmosphere temperatures measured by ascending thermistors on weather balloons show a decline of 0.36°C and satellites measuring the same layer (our only truly global measure) showed a decline of 0.24°C.

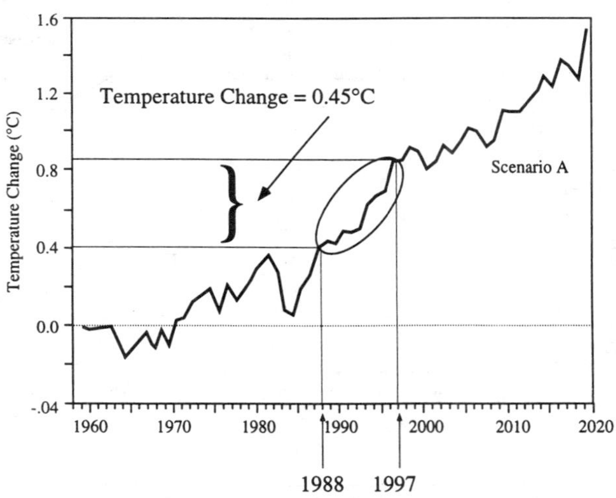

Figure 1. Hansen's global temperature projections from his 1988 model (Hansen et al., 1988.)

The forecast made in 1988 was an astounding failure, and IPCC's 1990 statement about the realistic nature of these projections was simply wrong.

This failure did not surprise me. On a 100 year time scale, the models were predicting a warming of about 1.5° by 1988. The observed change was 0.5°C. That the models continued to fail in the last ten years at the rate that they were failing in the previous century was strong evidence for my original thesis. How much might we have saved, including the notorious Kyoto Protocol, if we had just listened to nature instead of a manmade computer?

By 1995, in its second full Assessment of climate change, the IPCC admitted the validity of its critics' position: "When increases in greenhouse gases only are taken into account...most [climate models] produce a greater mean warming than has been observed to date, unless a lower climate sensitivity [to the greenhouse effect] is used...There is growing evidence that increases in sulfate aerosols are partially counteracting the [warming] due to increases in greenhouse gases."

IPCC is presenting two alternative hypotheses: Either the base warming was simply overestimated, or, some other anthropogenerated emission is preventing the warming from being observed. IPCC omitted a third source for the error: Perhaps the greenhouse gases were not increasing at the projected rate.

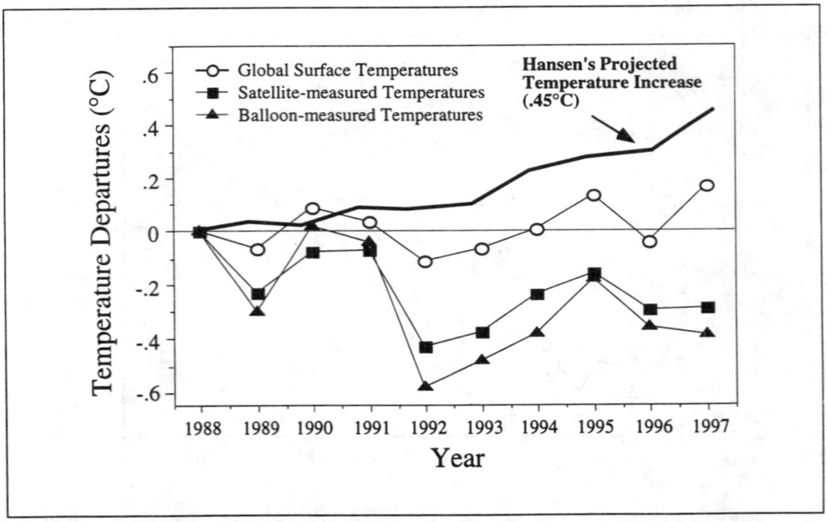

Figure 2. Hansen's global temperature projections from his 1988 model versus three independent observations of mean temperature during the period 1988 through 1997.

As evidence comes in, the first and third reasons appear to be carrying the day. The direct warming effect of carbon dioxide was overestimated (Myhre et al., 1998). Carbon dioxide is not accumulating in the atmosphere at even the lowest rate estimated by IPCC in 1992 (Hansen et al., 1998), and the the second most important greenhouse emission, methane, began to decrease its rate of increase in 1981 (Etheridge et al., 1998), some 15 years before the recent IPCC report that projects an *increased* rate of emissions for the next 50 years.

Only the sulfate hypothesis allows the exaggerated notion of climate change any credibility. It is not surprising that this is the one that IPCC continues to champion because it raises the spectre of "dangerous" interference in the climate system, which is what the Framework Convention on Climate Change was designed to prevent. If there is no "dangerous" interference, there is no need for the Convention, or the subsequent Kyoto Protocol, and the IPCC has failed in its mission. The U.N. General Assembly, more than ten years ago, directed the IPCC to provide the basis for the Convention.

Why did it not warm as predicted?

a. The sulfate hypothesis

Are sulfate aerosols responsible for the now-admitted dearth of warming? In previous testimony I have shown how poorly this argument stands the critical test of the data. Suffice it to say that the entire record of three dimensional atmospheric temperature does not appear consistent with this hypothesis. Instead of repeating that argument, I would simply point out that the southern half of the planet is virtually devoid of sulfates, and should have warmed at a prodigious and consistent rate for the last two decades. Unfortunately, we have very few longterm weather records from that half of the planet, and almost all come from the relatively uncommon landmasses. However, we do have nearly two decades of satellite data (Figure 3). They show a statistically significant decline in temperature—exactly the opposite to what the sulfate hypothesis predicts.

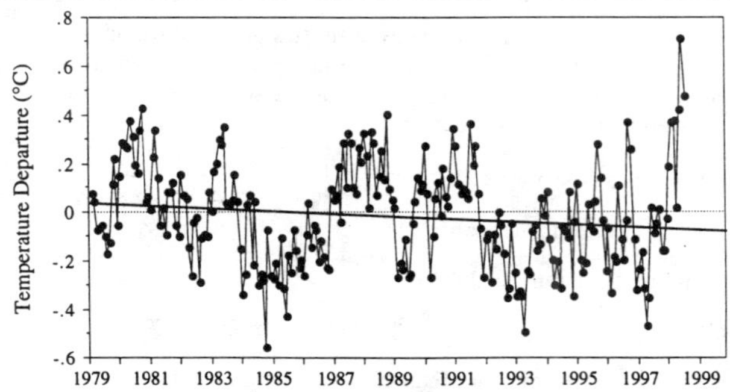

Figure 3. Satellite measured temperatures for the Southern Hemisphere show a statistically significant decline since records began in 1979.

b. Was the sensitivity overestimated?

If sulfates do not explain the lack of warming, one option is that the sensitivity to climate change was overestimated. The large warmings predicted by the failed models that back the Framework Convention rely on a roughly threefold amplification of carbon dioxide warming by increased atmospheric moisture. Yet Spencer and Braswell (1997) have found that the expected moisture is not there.

Perhaps even more remarkable is that amount of direct warming by carbon dioxide was also overestimated (Myhre et al., 1998). *This is the basic driving force behind the entire issue!*

c. Was the increase in greenhouse gases overestimated?

Dlugokencky et al. (1998) recently demonstrated that the concentration of methane in the atmosphere—currently 30% of the human greenhouse potential—is rapidly stabilizing. It has done this because its concentration is coming into chemical equilibrium with other atmospheric reactants. His calculations strongly suggest that the concentration will remain stable in the future. The IPCC assumed that, without any controls, the methane warming effect would *double* by 2050 and increase by 125% by 2100.

Hansen et al. (1998) recently calculated that the concentrations of carbon dioxide in the atmosphere are increasing at approximately 60% of the rate that is normally projected. Notably, he argues that the biosphere is absorbing CO_2 at a rate much faster than anticipated, as he wrote that "Apparently the rate of uptake by CO_2 sinks, either the ocean, or, *more likely the forests and soils* (our emphasis) has increased."

DECLINING PROJECTIONS OF GLOBAL WARMING

In the ten years since my first testimony, estimates of global warming to the year 2100 have declined. When the latest findings are factored in, the projected warming is now at the lower limit I noted in 1989. Following is a summary of that decline in median projected warming for the next century:

IPCC 1990 initial estimate: **3.2°C**

IPCC revised 1992 estimate: **2.6°C**

IPCC revised 1995 estimate: **2.0°C**

After allowing for overestimation of direct CO_2 warming: **1.7°C**

After allowing for flattening of Methane concentration: **1.4°C**

After allowing for decrease in carbon dioxide accumulation: **1.0°C**

The Nature of Observed Change

Winter Warming

Greenhouse physics predicts that the driest airmasses should respond first and most strongly to changes induced by human activities. These, in fact, are generally the coldest airmasses, such as the great high pressure system that dominates Siberia in the winter, and its only slightly more benign cousin in northwestern North America. When the jet stream attains a proper orientation, it is this airmass that migrates south and kills orange trees in Florida.

A look at the trends in the satellite data—our only truly global record of lower atmosphere temperature—is remarkably revealing. While there is no overall global warming trend, there is a pronounced warming trend in the coldest winter regions.

Balling, Michaels, et al. (1998) examined surface temperature records since 1945 and found also that warming was largely confined to the coldest winter airmasses, in agreement with the satellite. A warming of the coldest, driest airmasses, is by definition, a relative warming of the nights compared to the days. And, by extension, this is the type of climate change that slightly lengthens the growing season, as the coldest temperature occurs at night.

Climate Variability

Michaels et al. (1998) recently examined the surface temperature history in order to answer three questions:

Is the temperature becoming more variable from year-to-year? We found a statistically significant decline in interannual variability worldwide (Figure 4).

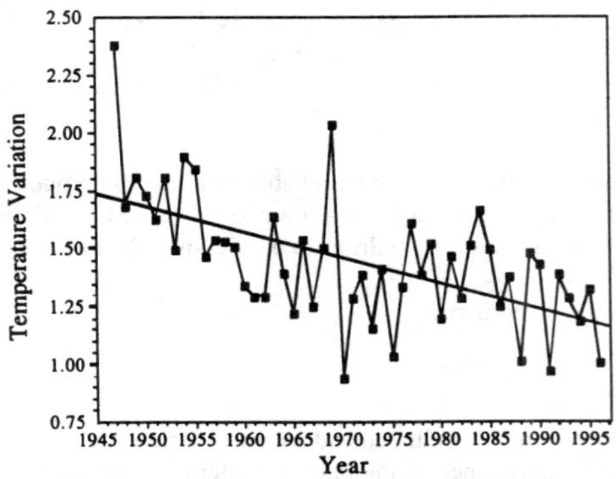

Figure 4. Annual temperature variation is declining, not increasing, on a global scale. From Michaels et al., 1998.

Is the variation from day-to-day increasing? We found no statistically significant change.

Are the number of record high or low temperatures increasing? We found no statistically significant change.

In summary, here is what the climate has done during the greenhouse enhancement: The most notable change is that the coldest airmasses of winter in Siberia and North America have warmed slightly. The only change in weather variability has been a tendency towards reduced year-to-year variability.

Our results should be integrated with a recent study of U.S. streamflow by Lins and Slack (1997). In an investigation of undisturbed sites, they found *no change* in the frequency of highest flow (flood) events, but a decrease in the lowest flow (drought) events.

We are not entering a world of increased variability, unpredictability and peril, but rather the opposite. If this is a human interference in the climate, it is hardly "dangerous."

The Kyoto Protocol: How Much Warming is Prevented?

This analysis assumes the IPCC's "consensus" estimate of 2.0°C of warming by the year 2100 in the absence of substantial emissions stabilization. Please note that my testimony indicates this is a considerable overestimation.

The Kyoto Protocol requires that the United States reduce its overall greenhouse gas emissions by a remarkable 43% for the 2008-2012 average, compared to where they would have been if we continue on the trajectory established in the last two decades. The economic costs are enormous, they are but not the subject of this hearing. What are the climate benefits?

Wigley (1998) recently calculated the "saved" warming, under the assumptions noted above, that would accrue if *every* nation met its obligations under the Kyoto Protocol. According to him, the earth's temperature in 2050 will be 0.07°C lower as a result. My own calculations produced a similar answer. Wigley is a Senior Scientist at the U.S. National Center for Atmospheric Research.

0.07°C is an amount so small that it cannot be reliably measured by ground-based thermometers. If one assumes the more likely scenario that warming to the year 2100 will be approximately half of the IPCC estimate, the saved warming drops to 0.04°C over the next fifty years.

This is no benefit at an enormous cost.

In conclusion, the observed data on climate and recent emissions trends clearly indicate that the concept of "dangerous" interference in the climate system is outmoded within any reasonable horizon. This makes the Kyoto Protocol a useless appendage to an irrelevant treaty. It is time to reconsider the Framework Convention.

References

Balling, R.C., Jr., et al., 1998. *Cli. Res.*, **9**, 175-181.

Dlugokencky, E.J., et al., 1998. *Nature*, **393**, 447-450.

Etheridge, D.M., et al., 1998. *J. Geophys. Res.*, **103**, 15979-15995.

Hansen, J.E., et al., 1988. *J. Geophys. Res.*, **93**, 9341-9364.

Hansen, J.E., et al., 1998. *Proc. Nat. Acad. Sci.*, **95**, 4113-4120.

Intergovernmental Panel on Climate Change, 1990. *Climate Change: The IPCC Scientific Assessment.*

_____, 1996. *The Science of Climate Change.*

Lins, H. and J.R. Slack, 1997. Amer. Geophys. Union, San Francisco. Abstract. 12/8/97.

Michaels, P.J., et al., 1998. *Cli. Res.*, **10**, 27-33.

Myhre, G., et al., 1998. *Geophys. Res. Lett.*, **25**, 2715-2718.

Spencer, R.W. and W.D. Braswell, 1997. *Bull. Amer. Met. Soc.*, **78**, 1097-1106.

Wigley, T.M.L., 1998. *Geophys. Res. Lett.*, **25**, 2285-2288.

Testimony of Dr. S. Fred Singer, President, SEPP to the Committee on Small Business, US House of Representatives (James M. Talent, Chairman)

2360 Rayburn House Office Building, July 27, 1998

Mr. Chairman and members of the Committee on Small Business:

Climate science does not support the Kyoto Protocol and its emission controls on carbon dioxide. As will become apparent from my testimony, the climate is not warming and climate models used to predict a future warming have not been validated. In any case, a warmer climate would be generally beneficial for agriculture and other human activities. Finally, as I will show, the Climate Treaty does not have a defined goal and does not provide any guidance on what level of atmospheric carbon dioxide is desirable.

I am an atmospheric scientist, with a Ph.D. in physics from Princeton University. I am now professor emeritus of environmental sciences at the University of Virginia and the president of the Fairfax-based Science and Environment Policy Project (SEPP), a non-profit educational and research organization. I have held several academic and government positions. The most relevant ones are as the first Director of the US Weather Satellite Service (which is now in the Department of Commerce) and, more recently, as a member and vice chairman of the National Advisory Committee on Oceans and Atmosphere (NACOA).

I have been involved with the subject of climate change since 1968 when I organized the first major symposium on "Global Effects of Environmental Pollution." I subsequently published an edited book by this title, followed by books on "The Changing Global Environment" and "Global Climate Change." My most recent book is "Hot Talk, Cold Science: Global Warming's Unfinished Debate," published in 1997 by The Independent Institute, Oakland, CA. My attached written testimony contains excerpts from this book, as well as from the SEPP policy pamphlet entitled "The Scientific Case Against the Climate Treaty."

The noted political scientist Aaron Wildavsky, who was a founding director of The Science & Environmental Policy Project, referred to global warming as "the mother of all environmental scares." This is a phenomenon which may not even be detectable; it has nevertheless been linked to all sorts of calamities: worldwide flooding due to rising sea levels, disastrous hurricanes

The Science & Environmental Policy Project
4084 University Drive, Suite 101, Fairfax, VA 22030-6812. Tel: (703) 934-6940. Fax: (703) 352-7535.

and floods, droughts and agricultural disasters, mass starvation, and the spread of tropical diseases putting three billion people at risk. Of course, none of these disasters are grounded in fact; they spring from the feverish imagination of activists and their ideological desire to impose controls on energy use and to stop economic growth—or at least micromanage it.

With climate change at the top of the national and even international agenda it is vital to distinguish reality from hype. I therefore welcome the opportunity provided by this hearing to set the record straight. Fearmongers have falsely claimed that there is a "scientific consensus" supporting their catastrophic scenarios. The opposite is more nearly true. The case against panicky policy actions -- be they punitive taxes on energy use or more direct control measures -- can be encapsulated in four points:

• Model calculations that predict a major warming in the next century have not been validated by actual observations of the climate over the past century. More specifically, the accurate data from weather satellites available since 1979 show no increase in global average temperature and directly contradict the theoretical models and computer simulations.

• Based on solid historic evidence there is every reason to believe that a modest increase in temperature, should it come about, is beneficial for human existence on the planet. Further, agriculturists concur that the ongoing increase in atmospheric carbon dioxide speeds up plant growth. Empirical evidence indicates that a climate warming, from whatever cause, should lower sea level, not raise it. What one should fear most then is the inevitable return of the next ice age -- which is now overdue.

• Even if greenhouse warming were really here, and even if it were as "bad" as hyped, what can realistically be done about it? People in developing countries want to improve their standard of living, and clamor for refrigerators, air conditioners, and cars. Can we deny them their aspirations? Keep in mind that stabilization of atmospheric CO_2 concentration at the present level requires more than just stabilization of emission: it requires that emissions be cut by 60 to 80 percent -- worldwide! Stabilization at twice the pre-industrial level requires an emission cut of nearly 50 percent from 1990 emission levels. Such a policy would foreclose economic development for most of the world's populations and condemn them to poverty.

• As a matter of fact, Article 2 of the Climate Treaty, which spells out the ultimate goal, does not specify a numerical target level. It simply calls for "stabilization of greenhouse gas concentrations in the atmosphere at a level that would prevent dangerous anthropogenic interference with the climate system." That level, however, is nowhere defined; nor is its definition addressed by any of the UN scientific reports. In other words, we don't know whether a higher or lower level of CO_2 would be more or less dangerous to the climate system. My own review of the published literature shows greater climate stability during the present warm interglacial than during the recent Ice Age. One could therefore conclude that a higher level of CO_2 is more desirable.

In conclusion: In spite of the fact that science does not support the emission cuts called for in the Kyoto Protocol, I do not argue for complacency. Any human-induced change in environment must be carefully monitored and evaluated. In the meantime, however, commonsense "no-regrets" policies like cost-effective energy conservation and improved efficiency are in order, rather than hasty and economically damaging actions based on insufficient science.

Testimony of

Robert T. Watson
Chair of the Intergovernmental Panel on Climate Change

before the

Committee on Small Business
U.S. House of Representatives

July 27, 1998

Mr. Chairman and members of the committee, it is pleasure to be appear before you today to discuss an issue of critical importance to this and future generations: global climate change. My name is Robert T. Watson and I am testifying today as the chair of the Intergovernmental Panel on Climate Change (IPCC).

Mr. Chairman, you are to be commended on holding this hearing to examine the scientific basis for the Kyoto Protocol to the United Nations Framework Convention on Climate Change. My testimony briefly describes the current state of understanding of the Earth's climate system and the influence of human activities; the vulnerability of human health, ecological systems, and socio-economic sectors to climate change; and approaches to reduce emissions and enhance sinks.

Overview

The overwhelming majority of scientific experts recognize that scientific uncertainties exist, but still believe that human-induced climate change is inevitable. The question is not whether climate will change in response to human activities, but rather where (regional patterns), when (the rate of change) and by how much (magnitude). It is also clear that climate change will adversely effect human health (particularly vector-borne diseases), ecological systems (particularly forests and coral reefs), and socio-economic sectors, including agriculture, forestry, fisheries, water resources, and human settlements, with developing countries being the most vulnerable. These are the fundamental conclusions of a careful and objective analysis of all relevant scientific, technical and economic information by thousands of experts (including, climatologists, theoretical modellers, oceanographers, meteorologists, chemists, biologists, technologists, economists and social scientists), from academia, governments, industry and environmental organizations from around the world under the auspices of the United Nations

International Panel on Climate Change. The good news is, however, that the majority of energy experts believe that significant reductions in greenhouse gas emissions are technically feasible due to an extensive array of technologies and policy measures in the energy supply and energy demand sectors at little or no cost to society. In addition, changes in land-use practices can also reduce net carbon emissions cost-effectively.

Decision-makers should realize that the atmospheric residence time of carbon dioxide, the major anthropogenic greenhouse gas, is more than a century, which means that if policy formulation waits until all scientific uncertainties are resolved, and carbon dioxide and other greenhouse gases are responsible for changing the Earth's climate as projected by all climate models, the time to reverse the human-induced changes in climate and the resulting environmental damages, would not be years or decades, but centuries to millennia, even if all emissions of greenhouse gases were terminated, which is clearly not practical. Figure 1 shows how sea level is projected to increase almost linearly for many centuries even after the atmospheric concentration of carbon dioxide is stabilized at twice today's level.

The Earth's Climate System: The Influence of Human Activities

The Earth's climate has been relatively stable (global temperature changes of less than 1°C over a century) during the present interglacial (i.e., the past 10,000 years). During this time modern society has evolved, and, in many cases, successfully adapted to the prevailing local climate and its natural variability. However, the Earth's climate is now changing. The Earth's surface temperature this century is as warm or warmer than any other century during the last six hundred years; the Earth's surface temperature has increased by about half a degree centigrade over the last century, with the land warming more than the oceans; the last few decades have been the hottest this century; and the five warmest years since instrumental records were established in 1860 were all in the 1990s, and eight of the ten next warmest years were in the 1980s and 1990s. 1997 was the warmest year on record and the first six months of this year are even warmer[1]. In addition, there is evidence of changes in sea level, that glaciers are retreating world-wide, and that the incidence of extreme weather events is increasing in some parts of the world. Not only is there evidence of a change in climate at the global level, but there is observational evidence that the climate of the U.S. is changing: increased temperatures (day and night), more intense rainfall events (defined as two inches within a 24 hour period), increased precipitation in winter, and less day-day variability in temperature.

---

[1] A significant amount of attention has been paid to the apparent discrepancy between temperature trends derived from surface observations made by thermometers, and satellite observations of the middle troposphere since 1979. It is important to note that there are a number of factors that may explain this discrepancy: (i) the observations are not measuring the same quantity, i.e., surface versus mid-troposphere; (ii) natural phenomena, such as El-Nino and volcanic eruptions could affect the data sets differently; and (iii) orbital drift and inter-satellite calibrations may affect the accuracy of the satellite-derived trends. In addition, it is important to remember that short-term (decadel) natural fluctuations are expected, hence the key issue is the trend in surface temperatures over the longer term that averages out these short-term natural fluctuations.

The atmospheric concentrations of greenhouse gases have increased because of human activities, primarily due to the combustion of fossil fuels (coal, oil and gas), deforestation and agricultural practices, since the beginning of the pre-industrial era around 1750: carbon dioxide by about 30%, methane by more than a factor of two, and nitrous oxide by about 15%. Their concentrations are higher now than at any time during the last 160,000 years, the period for which there are reliable ice-core data, and probably significantly longer. In addition, the atmospheric concentrations of sulfate aerosols have also increased. Greenhouse gases tend to warm the atmosphere and, in some regions, primarily in the Northern hemisphere, aerosols tend to cool the atmosphere.

Theoretical models that take into account the observed increases in the atmospheric concentrations of greenhouse gases and sulfate aerosols simulate the observed changes in surface temperature and the vertical distribution of temperature quite well. This, and other information, suggests that human activities are implicated in the observed changes in the Earth's climate. In fact, the observed changes in climate cannot be explained by natural phenomena alone (e.g., changes in solar output and volcanic emissions).

Future emissions of greenhouse gases and the sulfate aerosol precursor, sulfur dioxide, are sensitive to changes in population and gross domestic product, the rate of diffusion of new technologies into the market place, production and consumption patterns, land-use practices, energy intensity, and the price and availability of energy. Most projections suggest that greenhouse gas concentrations will increase significantly during the next century in the absence of policies specifically designed to address the issue of climate change. For example, carbon dioxide emissions from the combustion of fossil fuels are projected to range from 6 to 36 GtC per year in the year 2100: compared to current emissions of 6 GtC per year. However, in the last few months, two major oil companies, Shell and British Petroleum, have suggested that the mix of energy sources is likely to change radically during the next 50 years, with renewable energy sources (solar, wind and modern biomass) possibly accounting for as much as half of all energy produced by the middle of the next century. Such a future would clearly eliminate the highest projections of greenhouse gases being realized.

Based on the range of climate sensitivities (an increase in the equilibrium global mean surface temperature of 1.5 - 4.5°C for a doubling of atmospheric carbon dioxide concentrations) and plausible ranges of greenhouse gas and sulfur dioxide emissions (IPCC IS 92), climate models project that the global mean surface temperature could increase by 1 to 3.5°C by 2100. These projected global-average temperature changes would be greater than recent natural fluctuations and would also occur at a rate significantly faster than observed changes over the last 10,000 years. These long-term, large-scale, human-induced changes are likely to interact with natural climate variability on time-scales of days to decades (e.g., the El Nino-Southern Oscillation (ENSO) phenomena). Temperature changes are expected to differ by region with high latitudes projected to warm more than the global average. However, the reliability of regional scale predictions is still low. Associated with these estimated changes in temperature, sea level is projected to increase by 15 - 95 cm by 2100, caused primarily by thermal expansion of the oceans and the melting of glaciers. However, it should be noted that even when the atmospheric concentrations of greenhouse gases are stabilized, temperatures will continue to increase for several decades because of the thermal inertia of the climate system (temperature by another 30-50%), and sea level for an even longer period of time.

Model calculations show that evaporation will be enhanced as the climate warms, and that there will be an increase in global mean precipitation and an increase in the frequency of intense rainfall. However, not all land regions will experience an increase in precipitation, and even those land regions with increased precipitation may experience decreases in soil moisture, because of enhanced evaporation. Seasonal shifts in precipitation are also projected. In general, precipitation is projected to increase at high latitudes in winter, and soil moisture is projected to decrease in some mid-latitude continental regions during summer.

While the incidence of extreme temperature events, floods, droughts, soil moisture deficits, fires and pest outbreaks is expected to increase in some regions, it is unclear whether there will be changes in the frequency and intensity of extreme weather events such as tropical storms, cyclones, and tornadoes.

The Vulnerability of Ecological systems, Socio-economic Sectors and Human Health to Climate Change

The IPCC has assessed the potential consequences of changes in climate for human health, ecological systems and socio-economic sectors for ten continental- or subcontinental-scale regions: Africa, Australasia, Europe, Latin America, Middle East and Arid Asia, North America, Polar regions, Small Island States, Temperate Asia, and Tropical Asia. Because of uncertainties associated regional projections of climate change, the IPCC assessed the vulnerability of these natural and social systems to changes in climate, rather than attempting to provide quantitative predictions of the impacts of climate change at the regional level. Vulnerability is defined as the extent to which a natural or social system is susceptible to sustaining damage from climate change, and is a function of the sensitivity of a system to changes in climate and the ability to adapt the system to changes in climate. Hence a highly vulnerable system is one that is highly sensitive to modest changes in climate and one for which the ability to adapt is severely constrained.

Most impact studies have assessed how systems would respond to a climate change resulting from an arbitrary doubling of atmospheric carbon dioxide concentrations. Very few have considered the dynamic responses to steadily increasing greenhouse gas concentrations; fewer yet have been able to examine the consequences of increases beyond a doubling of greenhouse gas concentrations or to assess the implications of multiple stress factors.

The IPCC concluded that human health, terrestrial and aquatic ecological systems, and socioeconomic systems (e.g., agriculture, forestry, fisheries, water resources, and human settlements), which are all vital to human development and well-being, are all vulnerable to changes in climate, including the magnitude and rate of climate change, as well as to changes in climate variability. Whereas many regions are likely to experience the adverse effects of climate change—some of which are potentially irreversible—some effects of climate change are likely to be beneficial. Hence, different segments of society can expect to confront a variety of changes and the need to adapt to them.

There are a number of general conclusions that can be easily drawn: (i) human-induced climate change is an important new stress, particularly on ecological and socio-economic systems that are already affected by pollution, increasing resource demands, and non-sustainable management practices; (ii) the most vulnerable systems are those with the greatest sensitivity to climate change and the least adaptability; (iii) most systems are sensitive to both the magnitude and rate of climate change; (iv) many of the impacts are difficult to quantify because existing studies are limited in scope; and (v) successful adaptation depends upon technological advances, institutional arrangements, availability of financing and information exchange, and that vulnerability increases as adaptation capacity decreases. Therefore, developing countries are more vulnerable to climate change than developed countries.

The range of adaptation options for managed systems such as agriculture and water supply is generally increasing because of technological advances. However, some regions of the world, i.e., developing countries, have limited access to these technologies and appropriate information. The efficacy and cost-effectiveness of adaptation strategies will depend upon cultural, educational, managerial, institutional, legal and regulatory practices that are both domestic and international in scope. Incorporation of climate change concerns into resource-use and development decisions and plans for regularly scheduled investments in infrastructure will facilitate adaptation.

Let me now briefly discuss the implications of climate change for a representative number of systems: natural ecosystems (forests and coral reefs), food security, water resources, sea level rise, and human health.

Natural Ecosystems–Forests

The composition and geographic distribution of many ecosystems will shift as individual species respond to changes in climate, and there will likely be reductions in biological diversity (particularly species diversity) and in the goods and services ecosystems provide society, e.g., sources of food, fiber, medicines, recreation and tourism, and ecological services such as controlling nutrient cycling, waste quality, water run-off, and soil erosion. Models project that as a consequence of possible changes in temperature and water availability under doubled carbon dioxide equilibrium conditions, a substantial fraction (a global average of one-third, varying by region from one-seventh in tropical forests to two-thirds in Boreal forests) of the existing forested area of the world will undergo major changes in broad vegetation types. Climate change is expected to occur at a rapid rate relative to the speed at which forest species grow, reproduce and re-establish themselves. For mid-latitude regions a global average warming of 1-3.5°C over the next 100 years would be equivalent to a poleward shift of isotherms of approximately 150-550 km or an altitude shift of 150-550 meters. This compares to past tree species migration rates that are believed to be on the order of 4 - 200 km per century. Therefore, species composition of impacted forests is likely to change, entire forest types may disappear, while new assemblages of species and hence new forest ecosystems may be established. Large amounts of carbon could be released into the atmosphere during times of high forest mortality prior to regrowth of a mature forest.

Ecological models suggest significant changes in vegetation cover within the U.S. Potential changes include a northward shift in forest and other vegetation species, e.g., there would be an almost complete loss of sugar maple and beach trees in the Eastern US.

Natural Ecosystems--Coral Reefs

Coral reefs, the most biologically diverse marine ecosystems, are important for fisheries, tourism, coastal protection, and erosion control. Coral reef systems, which are already being threatened by pollution, unsustainable tourism and fishing practices, are very vulnerable to changes in climate. While these systems may be able to adapt to the projected increases in sea level, sustained increases in water temperatures of 3-4°C above long-term average seasonal maxima over a 6-month period can cause significant coral mortality; short-term increases on the order of only 1-2°C can cause "bleaching", leading to reef destruction. Indications are that the full restoration of coral communities could require several centuries.

Food Security

Currently, 800 million people are malnourished; as the world's population increases and incomes in some countries rise, food consumption is expected to double over the next three to four decades. Studies show that on the whole, global agricultural production could be maintained relative to baseline production in the face of climate change under doubled carbon dioxide equilibrium conditions. However, crop yields and changes in productivity due to climate change will vary considerably across regions and among localities, thus changing the patterns of production. In general, productivity is projected to increase in middle to high latitudes, depending on crop type, growing season, changes in temperature regime, and seasonally of precipitation, where-as in the tropics and subtropics, where some crops are near their maximum temperature tolerance and where dryland, non-irrigated agriculture dominates, yields are likely to decrease, especially in Africa and Latin America, where decreases in overall agricultural productivity of 30% are projected under doubled carbon dioxide conditions. Therefore there may be increased risk of hunger in some locations in the tropics and subtropics where many of the world's poorest people live.

While the productivity of agriculture in North America is moderately to highly sensitive to climate change, the vulnerability is thought to be low at the continental scale, although sub-regional variation losses or gains are likely. For example, a warming of 4-5 degrees Centigrade is projected to lead to negative impacts in Eastern, Southeastern and corn belt regions, but positive effects in northern plains and Western regions. These model calculations have considered the positive effects of higher atmospheric levels of carbon dioxide, but have not fully considered the effects of potential changes in climate variability, water availability, stresses from pests, diseases, and interactions with other existing stresses.

Water Resources

Currently 1.3 billion people do not have access to adequate supplies of safe water, and 2 billion people do not have access to adequate sanitation. Today, some nineteen countries, primarily in the Middle East and Africa, are classified as water-scarce or water-stressed. Even in the absence of climate change, this number is expected to double by 2025, in large part because of increases in demand from economic and population growth. Climate change will further exacerbate the frequency and magnitude of droughts in some places, in particular Africa where droughts are already a recurrent feature. Developing countries are highly vulnerable to climate change because many are located in arid and semi-arid areas.

The Great Plains and prairie regions of the US are particularly vulnerable, with projected increases in run-off in winter and spring and decreases in soil moisture and run-off in summer.

Sea Level Rise

Sea-level rise can have negative impacts on tourism, freshwater supplies, fisheries, exposed infrastructure, agricultural and dry lands, and wetlands. It is currently estimated that about half of the world's population lives in coastal zones, although there is a large variation among countries. Changes in climate will affect coastal systems through sea-level rise and an increase in storm-surge hazards, and possible changes in the frequency and/or intensity of extreme events. Impacts may vary across regions, and societal costs will greatly depend upon the vulnerability of the coastal system and the economic situation of the country. Sea-level rise will increase the vulnerability of coastal populations to flooding. An average of about 46 million people per year currently experience flooding due to storm surges; a 50 cm sea-level rise would increase this number to about 92 million; a 1 meter sea-level rise would increase this number to 118 million. The estimates will be substantially higher if one incorporates population growth projections. A number of studies have shown that small islands and deltaic areas are particularly vulnerable to a one-meter sea-level rise. In the absence of mitigation actions (e.g., building sea walls), land losses are projected to range from 1.0% for Egypt, 6% for Netherlands, 17.5% for Bangladesh, to about 80% of the Marshall Islands, displacing tens of millions of people, and in the case of low-lying Small Island States, the possible loss of whole cultures. Many nations face lost capital value in excess of 10% of GDP. While annual adaptation/protection costs for most of these nations are relatively modest (about 0.1% GDP), average annual costs to many small island states are much higher, several percent of GDP, assuming adaptation is possible.

Sea level rise is not only an issue for Small Island States and deltaic areas in developing countries, but poses a threat to many low-lying areas in the USA, in particular, the East coast and Gulf coast. IPCC noted that sea level has been rising relative to the land along most of the coast of North America, and falling in a few areas, for thousands of years. During the next century, a 50-cm rise in sea level from climate change alone could inundate 8,500 to 19,000 square kms of dry land, expand the 100 year flood plain by more than 23,000 square kms, and eliminate as much as 50% of North America's coastal wetlands. The projected changes in sea level due to climate change alone would under-estimate the total change in sea level from all causes along the eastern seaboard and Gulf coast of North America.

Human Health

Human health is sensitive to changes in climate because of changes in food security, water supply and quality, and the distribution of ecological systems. These impacts would be mostly adverse, and in many cases would cause some loss of life. Direct health effects would include increases in heat-related mortality and illness resulting from an anticipated increase in heatwaves. Indirect effects would include extensions of the range and season for vector organisms, thus increasing the transmission of vector-borne infectious diseases (e.g., malaria, dengue, yellow fever and encephalitis). Projected changes in climate under doubled carbon dioxide equilibrium conditions could lead to potential increases in malaria incidence of the order of 50–80 million additional cases annually, primarily in tropical, subtropical, and less well-protected temperate-zone populations. Some increases in non-vector-borne infectious diseases such as salmonellosis, cholera and other food- and water-related infections could also occur, particularly in tropical and subtropical regions, because of climatic impacts on water distribution and temperature, and on micro-organism proliferation.

Social Costs of Climate Change

The range of estimates of economic damages caused by changes in climate are quite uncertain. Taking into account both market and non-market costs, IPCC reported a reduction in world GDP of 1.5-2.0% for a doubled carbon dioxide environment. This value was obtained by summing widely varying estimates of damages by sector, including socio-economic sectors (e.g., agriculture, forestry, fisheries), ecological systems, and human health. Nordhaus, conducted an "expert" survey which resulted in a range from 0 to 21% for loss of world GDP, with a mean value of 3.6% and a median value of 1.9%.

Losses in developing countries are estimated to be much higher than the world average, ranging from 5% to 9%. Alternate assumptions about the value of a statistical life could increase the estimate of economic damages in developing countries.

IPCC reported values for the marginal damage of one extra ton of carbon emitted ranging from $5 to $125. A value of $5 to $12 per ton of carbon is obtained using a 5% social rate of ume preference (discount rate). Lower discount rates increase this estimate, e.g. a 2% discount rate would increase this estimate by an order of magnitude.

Approaches to Reduce Emissions and Enhance Sinks

Significant reductions in net greenhouse gas emissions are technically, and often economically, feasible and can be achieved by utilizing an extensive array of technologies and policy measures that accelerate technology diffusion in the energy supply (more efficient conversion of fossil fuels; switching from high to low carbon fossil fuels; decarbonization of flue gases and fuels, coupled with carbon dioxide storage; increasing the use of nuclear energy; and increased use of modern renewable sources of energy (e.g., plantation biomass, micro-hydro, and solar), energy demand (industry, transportation, and residential/commercial buildings) and agricultural/forestry sectors (altered management of agricultural soils and rangelands, restoration of degraded agricultural lands and rangelands, slowing deforestation, natural forest generation, establishment of tree plantations, promoting agroforestry, and improving the quality of the diet of ruminants). By the year 2100, the world's commercial energy system will be replaced at least twice offering opportunities to change the energy system without premature retirement of capital stock. However, full technical potential is rarely achieved because of a lack of information and cultural, institutional, legal and economic barriers.

Policy instruments can be used to facilitate the penetration of lower carbon intensive technologies and modified consumption patterns. These policies include: energy pricing strategies (e.g., carbon taxes and reduced energy subsidies); reducing or removing other subsidies that increase greenhouse gas emissions (e.g., agricultural and transport subsidies); incentives such as provisions for accelerated depreciation and reduced costs for the consumer; tradable emissions permits (and joint implementation); voluntary programs and negotiated agreements with industry; utility demand-side management programs; regulatory programs including minimum energy efficiency standards; market pull and demonstration programs that stimulate the development and application of advanced technologies; and product labeling. The optimum mix of policies will vary from country to country; policies need to be tailored for local situations and developed through consultation with stakeholders.

Estimates of the costs of mitigating climate change should take into account secondary benefits of switching from a fossil fuel based economy to a lower-carbon intensity energy system. Secondary benefits include lower levels of local and regional pollution, including particulates, ozone and acid rain.

The Challenge of Stabilization

It is important to recognize what emissions limitations are required in order to stabilize the atmospheric concentrations of carbon dioxide at different levels. Figure 2 shows time-dependent emissions profiles for stabilization at different levels (350 to 750 ppmv) - these profiles represent the so-called "delayed action" scenarios, i.e., the ones which some economists believe to be more economically efficient. It is instructive to note that: (i) global emissions will have to depart from IPCC IS 92a (classically referred to as business-as-usual) within the next few decades to achieve any of these stabilization levels, and (ii) independent of the eventual stabilization level, global emissions in the long-term will have to be well below today's level. Examination of the 550ppmv profile, a stabilization level suggested by European governments, shows that global emissions will have to depart from business-as-usual between 2010 and 2015, and peak between 2030 and 2035. In order to deviate from business-as-usual by 2010 to 2015 would require actions to be taken now.

One of the questions most frequently asked since Kyoto is: Will the emission reduction proposals agreed in Kyoto for industrialized countries lead to a stabilization of greenhouse gas concentrations, and if so, at what level? The answer is clearly no. Stabilization of carbon dioxide, the major anthropogenic greenhouse gas, will require global emissions reductions. Stabilization will eventually require global emissions to be reduced to between 2 and 3 billion tons of Carbon per year, in contrast to today's global emissions from energy use alone of about 6 billion tons of Carbon per year.

Summary

Policymakers are faced with responding to the risks posed by anthropogenic emissions of greenhouse gases in the face of significant scientific uncertainties. They should consider these uncertainties in the context that climate-induced environmental changes cannot be reversed quickly, if at all, due to the long time scales (decades to millennia) associated with the climate system. Decisions taken during the next few years may limit the range of possible policy options in the future because high near-term emissions would require deeper reductions in the future to meet any given target concentration. Delaying action might reduce the overall costs of mitigation because of potential technological advances but could increase both the rate and the eventual magnitude of climate change, and hence the adaptation and damage costs.

Policymakers will have to decide to what degree they want to take precautionary measures by mitigating greenhouse gas emissions and enhancing the resilience of vulnerable systems by means of adaptation. Uncertainty does not mean that a nation or the world community cannot position itself better to cope with the broad range of possible climate changes or protect against potentially costly future outcomes. Delaying such measures may leave a nation or the world poorly prepared to deal with adverse changes and may increase the possibility of irreversible or very costly consequences. Options for adapting to change or mitigating change that can be justified for other reasons today (e.g., abatement of air and water pollution) and make society more flexible or resilient to anticipated adverse effects of climate change appear particularly desirable.

If, actions are not taken to reduce the projected increase in greenhouse gas emissions, the Earth's climate is projected to change at an unprecedented rate with adverse consequences for society, undermining the very foundation of sustainable development. Adaptive strategies to deal with this issue need to be developed, recognizing issues of equity and cost-effectiveness.

While there is no debate that protection of the climate system will eventually need all countries to limit their greenhouse gas emissions, the Framework Convention on Climate Change recognizes the principle of differentiated responsibilities, and also recognizes that developed countries and countries with economies in transition should take the lead in limiting their greenhouse gas emissions given the historical and current emissions of greenhouse gases, and their financial, technical and institutional capabilities. Current and historical emissions of greenhouse gases arise mainly from developed countries and countries with economies in transition, i.e., emissions in developing countries are much lower, both in absolute and per capita terms. Even though it is well recognized that emissions from developing countries are increasing rapidly due to increases in population and economic growth, and are likely to surpass those from developed countries within a few decades (absolute terms, not per-capita), their contribution to global warming will not equal that of developed countries until nearly 2100 because the climate system responds to the cumulative emissions of greenhouse gases over time not the instantaneous annual emissions.

It is also quite clear that increased energy services in developing countries are critical in order to alleviate poverty and underdevelopment, where 1.3 billion people live on less than $1 per day, 3 billion people live on less than $2 per day, and 2 billion people are without electricity. Hence the challenge is to assist developing countries expand their production and consumption of energy in the most efficient and environmentally benign manner. The agreement in Kyoto by Annex I countries, coupled with an increased commitment to energy R&D, should stimulate the market for energy efficient technologies and low-carbon emitting technologies, thus allowing the U.S. to meet it's energy needs in a more climate friendly manner, but also providing a large market in developing countries for U.S. exports. Financial instruments such as the Global Environment Facility and promoting market mechanisms such as emissions trading (Article 17 of the Kyoto Protocol) and joint implementation (Articles 6 and 12 of the Kyoto Protocol) can assist in this endeavor--trading can be a win for the environment, reduce costs for developed countries, and provide a transfer of cleaner technologies and a share of the cost-savings to developing countries and countries with economies in transition.

Mr. Chairman, members of the Committee, I appreciate the opportunity you have provided me to be able to discuss these issues with you today. Thank-you.

Rise in sea-level

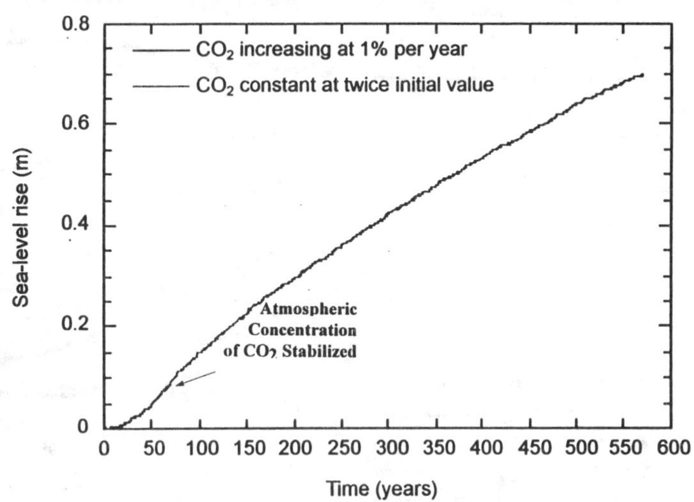

The rise in sea level due to thermal expansion alone (which accounts for about half of the total rise over the next decades), following an increase in atmospheric concentrations of CO_2 of 1% per year for 70 years (blue), and climate stabilization for a further 500 years (red).

The Met Office Hadley Centre for Climate Prediction and Research

FIGURE No. 2

Atmospheric Stabilization Emissions Paths

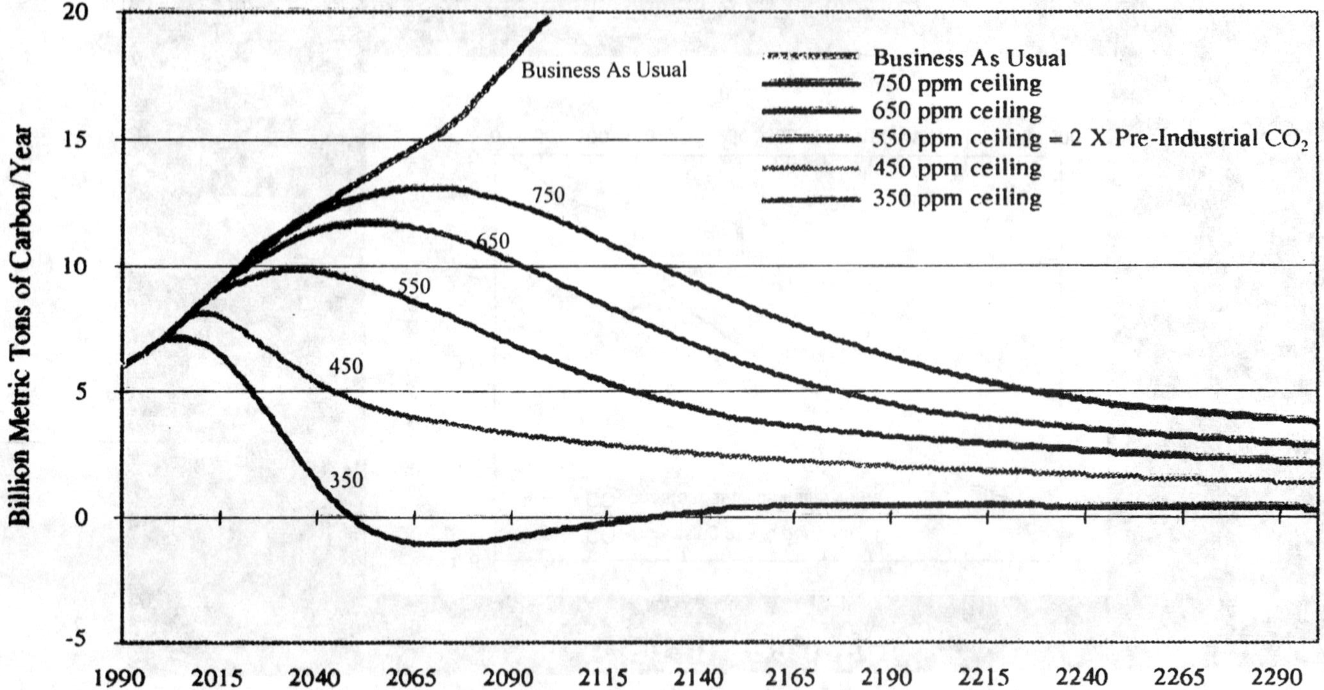

ROBERT T. WATSON

POSITIONS:
- Director, Environment Department, World Bank, Washington D.C.
- Chairman, World Meterological Organization-United Nations Environment Program Intergovernmental Panel on Climate Change (IPCC)

EXPERTISE :
Environmental Issues: establishing national and international research priorities; managing federal research programs; coordinating national and international research programs; coordinating international scientific assessments; linking scientific knowledge with policy formulation; and negotiating international environmental agreements

EDUCATION:
- Ph.D., (Gas Phase Chemical Kinetics) - 1973 — Queen Mary College, London University, London, England
- B.Sc. 1st class (chemistry) - 1969 — Queen Mary College, London University, London, England

PREVIOUS POSITIONS
- Senior Scientific Advisor, Environment Department, The World Bank, Washington, May 1996-July 1997
- Associate Director for Environment, Office of Science and Technology Policy, Old Executive Office Building, Washington, November 1973 - May 1996
- November 1993 - June 1996; Consultant, : Office of Science and Technology Policy, Old Executive Office Building, Washington D.C. 20506 - June 1993-November 1993
- Director, Science Division, and Chief Scientist (Acting), Office of Mission to Planet Earth
- National Aeronautics and Space Administration, Washington, D.C. April - November 1993
- Director, Process Studies Program Office, Earth Science and Applications Division, NASA – 1990-1993
- Branch Chief for Upper Atmospheric Research and Tropospheric Chemistry Programs, NASA – 1987-1990
- Acting Program Scientist (NASA) - 1980-1987 for (1) Upper Atmosphere Research Program; (2) Interdisciplinary Earth Science Research Program, and (3) Global Habitability Initiative — Detailee from the Jet Propulsion Laboratory, Pasadena, CA
- Research Scientist - 1980-1987; Member of the Technical Staff - 1978-1980; Senior Scientist - 1976-1978
- Postdoctoral Research Scientist
  University of Maryland, College Park, MD - 1974-1976; University of California, Berkeley, CA - 1973-1974

AWARDS:
- UNEP ozone protection award - 1995
- Environmental Protection Agency Award for scientific understanding of the ozone layer - 1994
- Climate Institute Scientific Achievement Award for Coordinating International Scientific Assessments - 1993

- American Association for the Advancement of Science Award for Scientific Freedom and Responsibility - 1993
- American Meteorological Society Special Award "for notable efforts in organizing and conducting international assessments in ozone depletion and global change" - 1993
- National Academy of Sciences Award for Scientific Reviewing - 1992
- American Geophysical Union's Edward A. Flinn Award - 1991
- Designated member of UNEP's "The Global 500: The Roll of Honour for Environmental Achievement" - 1989
- NASA Group Achievement Award (EOS) - 1993;
- NASA Cooperative External Achievement Award (High Speed Research Program) - 1992;
- NASA Distinguished Service Medal - 1989;
- NASA Public Service Group Achievement Award (Antarctic Aircraft Ozone Expedition) - 1989;
- NASA Group Achievement Award (International Ozone Trends Panel) - 1989;
- NASA OSSA - Edelson Award - 1987;
- NASA Public Service Group Achievement Award (International Ozone Assessment) - 1986;
- NASA Exceptional Service Medal - 1983

SPECIAL EXPERIENCE:

- Chair of the Science and Technical Advisory Panel to the Global Environmental Facility (World Bank, United Nations Development Program, and United Nations Environment Programme) 1991-1993
- Chair of an International Panel on the Scientific and policy Interlinkages Among the Global Environmental Issues - 1997-1999
- Co-chair: Intergovernmental Panel on Climate Change (IPCC) Working Group II - Impacts, Adaptation, and Mitigation, 1993- September 1997
- Co-chair: International WMO/UNEP Scientific Assessments of the Current Knowledge of the Processes Controlling the Abundance and Distribution of Atmospheric Ozone, 1985, 1988, 1989, 1991, 1994 and 1998
- Chair: UNEP/WMO Synthesis Panel for international Ozone Assessment (Science, Impacts, Technology, and Economics), 1989, 1991, 1994, 1998
- Chair: UNEP Global Biodiversity Assessment - 1993- 1996
- Member of the International Council of Scientific Unions (ICSU) Advisory Council on Environment - 1997 to present
- Chair: UNEP/WMO Synthesis Panel for Methyl Bromide (Science, Technology, and Economics) - 1992
- Representative on the Intergovernmental Panel on Climate Change (IPCC) Science Working Group and Member of Steering Group for IPCC. Chaired Chapter on Greenhouse Gases for the IPCC Scientific Assessment for Working Group I (1990 and 1992)
- Technical Advisor for the U.S. Government for both bilateral and multilateral negotiations of the Vienna Ozone Convention and the Montreal (CFC) Protocol
- Technical Advisor for the U.S. Government on Climate Change and member of the U.S. delegation to the International Negotiation Committee on Climate Change
- Co-chair: Committee on Earth and Environmental Sciences (CEES) Assessment Working Group
- Testified many times before U.S. Senate and U.S. House of Representatives Committees and Sub-Committees on Ozone, Global Warming, and other Global Change Research issues

PUBLICATIONS:
Thirty four peer-reviewed journal publications;
Thirty four major peer-reviewed scientific assessments and reviews

CONFERENCES/WORKSHOPS:
Hundreds of Keynote, Plenary and Invited talks

○